ABRAHAM GOTTLOB WERNER

NEW THEORY
OF THE
FORMATION OF VEINS
WITH ITS APPLICATION TO THE ART OF
WORKING MINES

Elibron Classics
www.elibron.com

Elibron Classics series.

ISBN 1-4021-6410-6 (paperback)
ISBN 1-4021-2162-8 (hardcover)

This Elibron Classics Replica Edition is an unabridged facsimile of the edition published in 1809 by the Encyclopædia Britannica Press, Edinburgh.

NEW THEORY

OF THE

FORMATION OF VEINS,

&c. &c.

E. Mitchell sculp.

WERNER

NEW THEORY

OF THE

FORMATION OF VEINS;

WITH ITS APPLICATION TO THE ART OF

WORKING MINES.

BY

ABRAHAM GOTTLOB WERNER,

COUNSELLOR OF THE MINES OF SAXONY, PROFESSOR OF MINERALOGY AND OF THE ART OF WORKING MINES AT FREYBERG, &c. &c.

TRANSLATED FROM THE GERMAN.

TO WHICH IS ADDED,

AN APPENDIX,

CONTAINING NOTES ILLUSTRATIVE OF THE SUBJECT;

BY CHARLES ANDERSON, M. D.

FELLOW OF THE ROYAL COLLEGE OF SURGEONS: MEMBER OF THE CHIRURGICAL SOCIETY, OF THE WERNERIAN NATURAL HISTORY SOCIETY, &c.

——————*Jam clarum mane fenestras*
Intrat, et angustas extendit lumine rimas. AUL. PERS. FLACCI SAT.

EDINBURGH:

Printed at the Encyclopædia Britannica Press;

FOR ARCHIBALD CONSTABLE AND COMPANY, EDINBURGH; AND CONSTABLE, HUNTER, PARK AND HUNTER, LONDON.

1809.

TO THE

PRESIDENT,

VICE-PRESIDENTS,

COUNCIL AND MEMBERS,

OF THE

WERNERIAN NATURAL HISTORY SOCIETY,

THE FOLLOWING PAGES ARE

MOST RESPECTFULLY

DEDICATED.

PERMIT me, Gentlemen, to present to the public, under the auspices of your name, the following translation of the celebrated Werner's Treatise on Veins, undertaken with a view of forwarding the

 object

object which first associated you together, the EXTENSION and PROMOTION of the study of NATURAL HISTORY.

I observe, with infinite satisfaction, the success which has already attended your exertions; and anticipate with pleasure the advantages which are likely to accrue from a continuation of them.

May your labours in that department of science be long useful; and may they be always as highly valued as I value them.

Accept of my best wishes for the prosperity of that institution, to which I am proud to have the honour of belonging; and believe me to be, with sentiments of the greatest respect,

Your most obedient

And very humble servant,

CHARLES ANDERSON.

LEITH, *April* 20. 1809.

TRANSLATOR'S PREFACE.

It is now upwards of seventeen years since this treatise, the only publication in which the celebrated Professor of Freyberg has given any exposition of the principles of his theory, was first presented to the public: yet strange as it may appear, it is no less true, that the doctrine it contains has been almost unknown, at least little studied, in this country, till within these very few years.

It would be no difficult matter to adduce a variety of reasons in explanation of this singular circumstance; but such a detail would be very uninteresting. Nothing perhaps has retarded a knowledge of the valuable observations and discoveries which it contains, more

than

than its having been written in German, a language so little cultivated in this country, that the perusal of this work has been confined to a few literary persons. For the same reason, much useful information on the nature of mines, and the mode of working them, is still as it were locked up from the use of the practical miners of this island. The number, variety, and importance of the mines in Germany, which annually yield an immense treasure, have created, in that part of the continent, an interest which is quite unknown here. Scientific men have there directed their attention to discover the manner in which ores and minerals occur in the bowels of the earth; and by reducing to fixed principles what they have been able to discover, they have, by their investigations, afforded much information, which has often shortened the labour of the practical miner, and been of much service in assisting him to solve many of the doubts and difficulties, which in his subterranean operations he frequently experiences.

It

It is a conviction of this, combined with a hope that it may be useful, which has induced the translator to present to his countrymen, in an English dress, the NEW THEORY OF THE FORMATION OF VEINS; a task which he undertook with much diffidence. How far his hopes of the utility of his labour may be realised, time only can determine. If, however, he shall be the means of giving more publicity to the doctrine which this treatise contains, by bringing it within the reach, not only of the scientific geognost, but of the practical miner who may not be able to peruse the original, the time he has bestowed on the translation will not have been employed in vain.

The translator is sensible that he has much reason to offer an apology for the manner in which he has executed his task. He is not insensible of the advantages which an ornamented style confers on any literary performance. But in the present instance it has been his ambition to render the original as closely as possible; trusting that the fidelity of the translation

tion will atone for any deficiencies which might strike the reader of cultivated taste.

Werner has a manner of expressing his sentiments which is peculiarly his own, and in many of the most valuable parts of his subject frequently makes use of a tautology which appears unnecessary. Of this the translator has in some instances ventured to retrench a little; though, in by far the greater number, he has not presumed to make any alteration. Wherever this has been done, he has preserved the sense of the original at the expence of giving to his sentence a turn which may often appear prolix, and which without this explanation would be thought inexcusable.

He has incorporated with the work a number of additions made by the author, which, however, do not appear in the original. These were given by Werner to M. Daubuisson, who inserted them in his translation into the French, and appear so valuable an addition, that to have omitted them would have been highly reprehensible.

The

The notes which he has ventured to add are such as seemed to afford some illustration of the particular parts of the doctrine which they are intended to explain; and though not so numerous as by some may be deemed necessary, they will, he anxiously hopes, be found to afford some information on the different topics of which they treat.

To conclude, the translator requests of all who, from their local situation or professional occupation, may have an opportunity of comparing the contents of this treatise with what they read in the great volume of nature, that they would correct and rectify their former observations by what they may here find, and note what on comparison they may observe not to correspond with the real phenomena of nature. We possess little or no information on the mining districts, or rather depots, of this country, for as yet they have not been divided into mining districts. That many important mineralogical facts may be collected from a careful survey of these mines cannot be doubted, ed, though we must regret that the attention of scientific men has not been directed to an examination of them conducted on just and fixed principles. To suppose that the proprietors and conductors of our mines do not possess the same inquisitive spirit, for which in other pursuits the natives of this kingdom are remarkable, would be to hazard a conjecture which will never be admitted. We must therefore seek some other cause to which this inattention is to be ascribed; and we cannot discover any which will more readily account for it, than the want of some just and correct observations founded on facts as they present themselves in exploring the great and hidden treasures which this country contains within its bowels.

On the principles of this treatise, Werner has reared that beautiful superstructure to account for the mode of formation of this earth in general, which has obtained his name. It is the result of 30 years close and unremitted application to the nature of veins and the mineral

neral masses which compose the crust of the globe, and is deduced from facts alone, to the exclusion of all hypothesis. Such a mode of philosophizing, viz. observing facts, arguing from them, and then establishing a theory on what has been seen and can be explained, gives us all the probability of reaching the truth that legitimate induction is able to afford. It is this which has conferred its superior excellence on the Wernerian theory, and procured for its author so great a reputation.

AUTHOR'S PREFACE.

THE uninterrupted geognostic observations in which I have been engaged, have been principally directed to a consideration of the different mineral repositories; and, with the view of rendering my remarks more useful and applicable to the practical operations of mining, my attention has been directed to every circumstance which could make me acquainted with the nature and peculiarities of each of these, but more particularly of veins, which, being of great variety, and having undergone many changes, are of a very complicated nature. In studying veins, I have not only applied myself to discover, as much as possible, their nature and properties, as well as their different relations to one another, and to the substance of the rocks in which they occur, but I have also

endeavoured

endeavoured to obtain some knowledge of their origin and mode of formation. From this source I have attempted to derive the explanation of their peculiarities, and of the various phenomena which they exhibit.

The discoveries which I have been fortunate enough to make on this subject, greatly surpass what my most sanguine wishes could have led me to expect, in a field of inquiry ſo intricate and complicated. I now present to mineralogists and miners the result of these discoveries, as well as the new theory of the formation of veins deduced from them. It will add to our knowledge of geognosy, principally in the natural history of our globe, and in the science of mine-working.

This new theory will fill up a considerable blank in geognosy; it will throw new light on the various revolutions which our globe has undergone, and it will open an ample field for future observations and enquiries. It will increase and confirm the connection which subsists between orictognosy and geognosy, and even between the science of mine-working and

orictognosy;

orictognosy; and it will afford opportunity for making new applications of this last branch of mineralogy.

All those who know how much, in mining, we are often obliged to grope in the dark, whether in seeking for a known or unknown vein which we wish to work, or to discover in a vein the points and places which contain the largest quantity of the metallic minerals, or to regain a vein which has already been worked, but has been lost, which very frequently happens; all such persons will easily appreciate the value of the new theory to the miner, in proportion as it is rendered more perfect, and is applied to the arrangements and works carried on in the interior of mines. In order to set this matter in the clearest point of view, and to give a more perfect idea of it, I have devoted a whole chapter to the detail of the many advantages to be derived from it.

Further, In the second chapter, where I give a history of former doctrines on this subject, I quote the opinion respecting my researches entertained by Mr Von Oppel, certainly an impartial person, who has been long celebrated among miners both for his knowledge and his experience. The exposition which I have made of the utility of my theory in the ninth chapter, and the opinion of one so competent to judge of it, will, I flatter myself, be sufficient to convince all well informed miners of its utility; such at least as are possessed of experience and knowledge, and shall study the subject without prejudice. I do not, however, rest satisfied with merely stating its utility, but I also show in detail the application of it to the working of mines. I also treat of the manner of attaining this object, by constructing maps, drawing up descriptions of mining districts, and forming collections of minerals; the whole being arranged in a proper manner. In the tenth chapter I apply this theory to the mining district of Freyberg, in every respect the most important in our mountains. This application consists of a description of the different formations of the metalliferous veins contained in the district. In fine, as miners may have adopted some of the ancient

cient theories, and as their attachment to these may form an obstacle to the advantages that are to be derived from the one I here give an account of; I have been under the necessity, in spite of my aversion to polemical discussions, of pointing out in the most evident manner the falsity and groundlessness of such theories. This I have done in the eighth chapter.

The theory which I detail in this treatise is founded upon observations and facts partly well known, partly new, but which can be easily verified; it is also founded upon incontestable and generally admitted theories respecting the laws and effects of nature; and lastly, upon analogies and consequences, to the exclusion of every thing hypothetical. As every one acquainted with the subject can judge of it, I am equally desirous of seeing it subjected to numerous and rigid trials, as of hearing that a great number of facts are collected to rectify and complete the observations brought forward to support and explain it.

I must here present a request to all who would judge of this theory, or communicate

their

their sentiments on it to the public, viz. to begin by first reading through the whole treatise, and then to peruse it a second time, with attention. Such a request will not only appear strange to many persons, but even superfluous; I find myself, however, under the necessity of making it, from the manner followed by some individuals with my book on the external characters of fossils. They have often represented me as saying quite the contrary of what I have expressly written. This may have been done, by some, through design, but in by far the greater number of instances, it has happened from that work not having been read through; and in this manner the public has frequently been led into an error, at least for some time.

The proposition, that the spaces now occupied by veins, were originally rents formed in the substance of rocks, is not a new opinion; it had formerly been propofed, and adopted by the greater number of geognosts, as will be feen in the chapter of this treatise where I give an abridged history of the ancient theories of veins.

I only claim the merit of having ascertained, in a more positive manner, the causes which have produced these rents, and of having brought forward better proofs of it than had formerly been done. But, what is altogether new, and what I may challenge as my own particular discovery is, *a.* to have determined, and described in a more particular manner, the internal structure of veins, as well as the formation of the different substances of which they are composed, and to have settled the relative age of each; *b.* to have given the most accurate observations and most perfect knowledge of the meetings and intersections of veins, and to have made these observations subservient to the determining their relative ages; *c.* to have determined the different vein-formations, particularly metalliferous veins, as well as their age; *d.* to have been the first who entertained the idea, that the spaces which veins occupy were filled by precipitations from the solutions, which at the same time formed, by other precipitations, the beds of mountains; and to have furnished proofs of this; and *e.* to have determined the essential differences that are found between the structure of veins and that of beds.

Farther, Almost all the refutations of objections brought against my theory, as well as those which I have urged against former ones, and the greater part of the applications made of it to the working of mines, are peculiar to myself. I shall say nothing of my claim to several other discoveries. For upwards of 17 years I have publicly discoursed of veins in my courses of geognosy, and in those also which I have delivered on the manner of working mines. From these many persons have imbibed my ideas, and communicated them to the public, either directly or indirectly; so that it is very difficult for me now to claim them as my own.

From my earliest youth I began to collect, in Lusatia and Silesia, the observations on which my theory is founded; but the greater number has been furnished by the mountains of Saxony, called Erzgebirge. During the 22 years that I have been settled in this country, I have often descended into the interior of the mines, in the district of Freyberg, and into all the others.

Besides,

Besides, I have derived much knowledge from journeys made in the mountains of Bohemia, chiefly at Joachimstahl; in the mountains too of Franconia, of Hesse, and of Thuringia. My limited fortune, and the nature of my present situation, have prevented me from travelling into more distant countries, however useful I may consider such journeys to be, and however desirous I may have been to undertake them. After this I cannot think that any person, who considers the observations I have made and collected on this subject, and on which I have grounded my theory, will upbraid me with having only known one single tract of country; this reproach has been once made, though erroneously, and on the subject of another part of geognosy. Independently of my own observations on veins, I have derived much assistance from the investigations furnished to me by the numerous pupils I have had from almost every part of Europe, or from those which I have obtained by an extensive correspondence carried on with a great many mineralogists and experienced miners; investigations which have confirmed the greater part of my

own

own observations. I have also used, though with caution, the observations made by other geognosts in different countries, which are to be found in their writings. Lastly, The attentive and strict examination of the numerous fossils and specimens of veins in my own cabinet, or of those I have seen in other collections, has been of great assistance to my theory; for in examining a piece of mineral, I am not satisfied with ascertaining, as is generally the case, what are the different fossils of which it is composed, but I endeavour to discover a certain formation of metalliferous veins, and to determine whether these fossils have been formed at the same time, or which of them are of newer and which of older formation than the rest.

Although I have been nearly 30 years occupied in studying the nature of veins, and in investigating their formation, and although it is now six years since I first explained, in my course of geognosy, the same theory that is given and demonstrated in this treatise, yet I have not bestowed all the pains in digesting it which I could have wished; consequently it is not

not divided and arranged throughout in the most proper manner. I have been under the necessity of composing and putting it in order within the space of three months; and that at a time too, when I have been engaged for eight or ten hours a day in works requiring great exertion of mind; every sheet was printed as soon as it was written, so that I had no opportunity of altering or revising it, far less of correcting the whole.

After having, in my capacity of author, given to the public that account of the work presented to them which they have a right to expect, I shall now make a request to all persons who are sincerely interested in the advancement of mineralogy and the science of mines; which is, that they second my efforts and assist me in bringing to perfection my New Theory of Veins, and in discovering new formations of veins, particularly such as are metalliferous, to transmit to me all the documents they can relative to them, accompanied with specimens of the fossils. In these documents I include among other things an orictognostic and correct description of the substance

substance of the veins; a particular notice of the remarkable peculiarities presented by certain veins relatively to one another, such as their joinings, intersections, &c.; complete descriptions of the veins; descriptions also of whole depots of ores, and of mining districts; remarks on the analogies and resemblances which certain vein-formations bear to one another, as well as to beds; and an account of any singular phenomena which may either serve to confirm or overturn my theory. I shall make use of every thing sent me on this subject, and shall receive with equal pleasure every thing addressed to me, the object of which shall be either to rectify my errors, or entirely to refute my opinions, provided it be done in a becoming manner.

I take this opportunity of pointing out an error into which some of my friends have fallen, which has in several respects been prejudicial to me, and is still likely to be so. Such persons have conceived a singular notion which they have publicly circulated, by saying that I have formed the resolution of publishing none of of my numerous and painful labours in mineralogy, the science of mines, forges and foundries. They think (I am persuaded in consesequence of the good opinion they have formed of me and my scientific researches), that they render an essential service to the sciences in publishing as soon as possible the result of my observations and labours.

This cannot be a difficult matter; for ever since I became a professor in this school, my manner of lecturing has been such that my hearers might take notes, or even write every word, under my own direction. In this manner several of my courses of lectures have been written; and to my great displeasure, a sort of mercantile speculation has been made of these manuscripts in other countries. They are for the most part defective, although some of the number are more perfect than others. On what is past, I am willing to shut my eyes; but as I understand there is an intention of publishing and printing not only the introductory part of my system of orictognosy, but also my dissertation on Iron Foundries, I think, in in justice to myself and the science, I am bound highly to disapprove of such an undertaking. For besides being thus robbed of my property as an author, the works presented to the public in this way cannot but be very defective, and in some degree mutilated. I here announce that I am engaged in revising my works on orictognosy, geognosy, and the other branches of the science, and that they shall forthwith appear one after another, enriched by my latest observations and discoveries.

(Signed) ABRAHAM GOTTLOB WERNER.

FREYBERG,
November 20. 1791.

CONTENTS.

CHAP. I. Of Veins in General.

CHAP. II. Short Account of the different Theories on the Formation of Veins.

Chap. III. Short Account of the New Theory of Veins, and of the Mode of their Formation.

§ 35.

CHAP. V. Farther Elucidation of the Proofs brought forward, and of the Theory resulting from them; and a Refutation of some Objections to it.

CHAP. VI. Proofs that the Spaces which Veins now occupy have been filled from above.

CHAP. VII.

CHAP. VII. Reply to the Objections which have been and may be brought against my theory, and continuation of the account of it, &c.

CHAP. VIII.

CHAP. VIII. A short Refutation of the Ancient Theories of the Formation of Veins.

CHAP. IX. Of the application of the New Theory of Veins to the working of Mines.

§ 97.

§ 117.

NEW THEORY

OF THE

FORMATION

OF

VEINS.

CHAP. I.

OF VEINS IN GENERAL.

§ 1.

To assist the reader in acquiring a more perfect knowledge of what I have to say on the formation of veins, I deem it necessary to commence by giving a short account of veins in general; next, I ſhall explain some new technical terms which I have found it neceſsary to make use of in this treatise; and then

I shall give a concise history of the former theories which have been proposed on this subject.

In this chapter I shall give the definition of the terms I employ; and the second will be occupied by an account of the old theories.

The first definition I shall give is that of a *vein*. In explaining this term, I shall mark the difference between what are to be considered *true veins*, and what are not so, even among various mineral repositories to which the name of veins and other appellations have been given by miners.

§ 2.

Veins are particular mineral repositories of a flat or tabular shape, which in general traverse the strata of mountains, and are filled with mineral matter differing more or less from the nature of the rocks in which they occur.

Strata are those parts of a rock which lie between parallel rents and fissures. These separated parts or strata are therefore tabular masses of the same nature and substance, parallel to each other, and of different degrees of thickness. Mountains and rocks thus divided are called *stratified rocks*: all rocks, however, are not stratified.

From the differences in the position, division, and

thickness

thickness of the strata, arise the differences of stratification, which is to be distinguished from the superposition of rocks. By *super-position* is to be understood the order in which the different rocks composing the crust of the globe are arranged above one another. Nor must we confound stratification with the structure of rocks, though nearly allied to each other, especially in rocks of a slaty structure, where the direction of the strata and folia are the same.

It will be a more accurate description of *veins*, to say that they are rents which have been formed in mountains, and have been afterwards filled up by mineral matter differing more or less from the nature of the rocks *.

§ 3.

Veins, as we have remarked, cross the strata, and have a direction different from theirs. Other mineral repositories

* I only add this last definition of veins to render the former more explicit: for in determining the formation of veins, it brings their distinctive character in a more striking manner under our observation. I did not wish to give it at first as a definition, because it takes for granted a formation which remains to be proved in the course of this treatise; and when this formation is proved, it may be used with the greatest propriety.

repositories (such as particular strata or beds, of whatever thickness they occur) have, on the contrary, a similar direction with the strata of the rock; and, instead of crossing, run parallel with them. This forms the characteristic difference. The Swedish mineralogists give the name of *gangar* to mineral repositories in general: veins, properly so called, they name *skiælard* *.

§ 4.

A *stockwerke*, in the real acceptation of the word, (for formerly it was applied to a certain manner of working a mine), is part of a rock of greater or less extent, which is penetrated and crossed in all directions by an almost innumerable quantity of small veins.

What the miners in Hungary call rents, *(klüfte)* are small and middle-sized veins. What, in floetz countries, are denominated by them *rükken* and *wechsel*, are true veins; the first of these is often of considerable magnitude.

The veins which traverse the rocks of the coal formation

* On this, consult Wallerii Elementa Metallurgiæ, § 14.

mation near Dresden, are improperly denominated *kamme*.

§ 5.

I suppose my readers are already acquainted with the technical terms employed in speaking of veins, and that they understand what is meant by the magnitude, position, and structure of a vein *. I must, however, explain some others which my new theory has rendered it necessary to introduce.

By a *vein formation*, or simply a *formation*, is to be understood all veins formed at the same time, and having one and the same origin; whether they occur near each other in the same country, or are met with in different and distant countries. In this manner I employ the terms *lead glance, fluor, and heavy spar formation; red and white silver ore formation.*

When several veins of the same formation occur in a country, they form what I call a *depot of veins*. This I distinguish by the name of the country in which

* On the different technical terms used in describing veins, as also on their position, magnitude, &c. different works on mining may be consulted, particularly the first part of the *Bericht vom Bergbau*, 4to. Freyberg 1749.

which it is found joined to the principal fossile which it contains. In this way I say the *lead glance, grey copper ore, and yellow blende depot of Scharfenberg: the tin depot of Altenberg.*

Lastly, By *mining district* is to be understood, many mineral deposits occurring in the same country, which generally cross each other. I distinguish such districts by the name of the country where they occur, as the *mining district of Freyberg.*

I shall now proceed to give the history of the different theories which have hitherto been proposed to account for the formation of veins.

CHAP.

CHAP. II.

A SHORT HISTORY OF THE DIFFERENT THEORIES ON THE FORMATION OF VEINS.

§ 6.

Among the Greek and Latin classics who treat of mines and minerals, we find very little on the subject of veins. It will appear, however, from some passages which I am to quote, that they were not altogether ignorant of the subject. There can be no doubt that veins were known long before the time of these authors. They must unquestionably have become an object of attention as soon as mines began to be worked. But it is not probable, that at that period, veins were distinguished from the other mineral repositories, or that any attempts were made to account for their mode of formation: at least such attempts are unknown to us.

Diodorus

Diodorus Siculus *, at the beginning of the passage where he speaks of the famous gold mines of Egypt, says, " the mountains in this country are " of a dark hue; they are traversed by veins of a " white stone, possessing great lustre; it is from this " ſtone that gold is extracted †." Farther on, when treating of the gold and silver mines of Spain, he says, " the mountains in that country are traversed " by many metallic veins ‡."

Pliny,

* Diodori Siculi Bibliothecæ Hiſtoricæ, libr. xv. per Laurent. Rhodomannum. Hanov. 1604. fol.

This author lived and wrote about the beginning of the Chriſtian æra.

† Τῆς γὰρ γῆς μελαίνης ὄυσης τῇ φύσει καὶ διαφυὰς καὶ φλέβας ἐχούσης μαρμάρου τῇ λευκότητι διαφερόυσας, καὶ πάσας τάς περιλαμπομένας φύσεις ὑπερβαλλὄσας τῇ λαμπρότητι, οἱ προσεδρεύοντες τοῖς μεταλλικοῖς ἔργοῖς τῷ πλήθει τῶν ἐργαζομένων κατασκευάζουσι τὸν χρυσόν.

The black mountain here mentioned is probably composed of dark clay slate, and the white marble is quartz or some kind of spar, most likely the former.

‡ Πᾶσα γὰρ ἡ σύνεγγις διαπέπλεκται πολυμερῶς τοῖς ἑλιγμοῖς τῶν γὰρ ῥάβδων.

The διαπέπλεκται πολυμερῶς, as well as διαφέρουσας, in the former passage, shows that the repositories in the mines of gold and silver frequently traversed the maſs of the rock in question in all directions, and that they were consequently true veins.

Pliny, also, in speaking of the manner in which gold is found in the mountains, says, "the veins of "gold run through the rocks in many different di- "rections, and traverse the walls of wells *."

§ 7.

Agricola †, the first of the moderns who has written on veins, has given a very particular account of

B them.

* Vagantur hi venarum canales per latera puteorum, et huc illuc (Caii Plinii Secundi Historiæ Natur. lib. xxxvii. quos interpret. et not. illustravit Jo. Harduinus Pausus, 1723, fol. lib. xxxiii. cap. xxi. tom. ii. p. 617.)

Under the name of *canales* is to be understood, not galleries, which are called *cuniculi* and *cryptæ*, but the veins themselves. In this sense the words have been considered by Agricola: in the same manner, the gold which is found in veins is called *aurum canalicium*, to distinguish it from that obtained by washing, and of which Pliny speaks before.

The word *vagari* means that this kind of repository extends in all directions in the rock, and that consequently they were true veins.

In the edition of Dalechamp, it is said, *hi venarum canales* per marmor *vagantur et latera puteorum et hic illuc*.

Pliny wrote his work about the year 100.

† George Agricola (whose proper name was Bauer,) was born in 1494 at Glaucha in the Saxon Erzgebirge. He

them. He speaks of them in many parts of his writings, and endeavours not only to define and describe them, but also to account for their mode of formation. In the whole of his researches on this difficult subject, he not only surpasses all who have preceded him in the same path, but even those who lived a century after him.—Nevertheless he was unacquainted with the characteristic differences between true veins and other mineral repositories. He treats of the size, position, and crossing of several veins, in the 24th chapter of his work entitled *Bermannus*, and still more particularly in the third book of his great work, *De Re Metallica* †.

In

He practised medicine first at Joachimstahl, and then at Chemnitz; at which last place he died in 1555. He is the father of mineralogy, and of the science of mining. To an intimate acquaintance with the sciences, he joined much classical learning. His Latin was so pure, that this circumstance alone is sufficient to entitle his works to a place amongst other classics. His writings were numerous; and, with the exception of a few tracts, have been published in two vols. fol. by Froben and Basel.

Agricola's reputation in the sciences which he cultivated, will last as long as Pliny's in natural history, or Aristotle's in philosophy.

† Georgii Agricola De Re Metallica, libri xii. fol. Basilea, 1556.

In the third chapter of the third book of his work *De Ortu et Causis Subterraneorum* *, he treats of the formation

---

* This work, with several other writings of Agricola, and among the number his *Bermannus*, have been published in one volume, with the following titles.

De Ortu et Causis Subterraneorum libri v.

De Natura eorum quæ effluent ex Terra libri iv.

De Natura Fossilium libri x.

De Veteribus et Novis Metallis libri xi.

Bermannus, sive De Re Metallica Dialogus.

Interpretatio Germanica Vocum Rei Metallicæ, addito indice fœcundissimo.

The passages alluded to are the following.

" Sed commissuras saxorum duobus modis fieri intelligimus; uno qui proprius ipsarum est cum saxa gignuntur: tum enim materiam tenacem calor coquitur in lapidem: at non lenta, similiter cocta exhaurit humorem efficiturque terra plerumque friabilis. Altero, qui communis ipsis est cum fibris et venis, cum aqua in unum locum colligetur. Ea enim saxa mollit suo liquore, sua gravitate et pondere perrumpit et dividit: quod si dura fuerint, efficit commissuras saxorum et fibras; sin non nimis dura, venas." Agricola denominates rents *commissuras;* small veins and ramifications, *fibras;* large veins he calls *venas;* and the whole together *terræ canales*. For at the beginning of this chapter he says, " venio nunc ad terræ canales, hoc est, venas, fibras, et quas commissuras saxorum vocantur; quæ quidem vasa sunt aut receptacula materiæ ex qua res fossiles formantur."

This

formation of veins. The rents and fissures in which veins are found, he supposes to have been formed, partly at the same time with the rocks themselves, and partly afterwards, by the waters penetrating them; so that, where there has been either a large quantity of water, or where the substance of the rock has been much softened, there the largest fissures occur; where this has not happened, the fissures are proportionably smaller. With respect to the earths and stones which are found in veins, the former he conceives to have been detached from the rocks, and carried into the veins by the water *; the

This author, in treating of the formation of veins, in the same chapter, says; "Impetus aquarum saxa in quo-"que loco fragilia comminuit et diffindit. Igitur per ea "fracta et diffissa permeat et transit modo in profundum, "fibrasque vel venas facit profundas: tum autem in latum "quo modo dilatatæ fiunt."

By the conclusion of this passage it appears, that he comprehends beds as well as other repositories under the name of veins.

* This is obviously his opinion, from the two following passages of the work already quoted.

Lib. iii. chap. iv. "Terra pura, sive simplex, in canali-"bus hoc modo gignitur. Aqua pluvia, quam summum "terræ imbibit, primo per ejusdem terræ interiora per-"meat

the latter he considers as arising from the earthy matters hardened by the effects partly of heat and cold, and partly by a lapidific juice *. Minerals and metals

---

" meat et transit, cumque ipsa miscetur: deinde undique " colligitur in fibras et venas." A little farther on he adds, " aqua autem sic mista in unum aliquem canalis al- " veum confluit, vel in angustias abducta percolatur: quo- " modo materia pura et æqualis subsidet; aqua vero deficit " et delabitur. Ex qua sanè materia oritur terra, de qua " nunc loquimur."

* This same author says, in the following passages of the fore-mentioned work,

Lib. iv. cap. iv. " Albertus, Arabem (i. e. Avicen- " nam) secutus, scribit, materiam lapidum esse speciem " quandum terræ aut aquæ: id est, duo hæc, terram dico " et aquam, sic inter se mista, ut modo hoc, modo illæ " vincat alterum mole. Uterque (sc. eorum) autem quod " ita sensit, ratione fecit. Nam non fit lapis ex pura neque " terra, neque aqua: quod solam terram præ siccitate ejus " non coagmentet ac conglutinet calor, sed dissolvat magis " faciatque pulverem. Aqua vero simplex congelascit qui- " dem frigore, sed modicus tepor eam resolvit." He adds farther, " Humor vero vicissim congregationis conjunc- " tionisque partium rei siccæ, quoddam quasi vinculum " est." Then cap. v. " Itaque si permistio abundaverit " terra, discitur lutum: sin aqua, succus. Nihil enim " aliud est lutum, quam terra quæ permaduerit aqua, ut " superiore libro explicavi: nec succus aliud quiddam est, " quam aqua, quæ contra sorbuerit terram, vel corroserit " tetigeritque

metals he regards as being deposited from a solution in water, containing the earthy materials intimately mixed and combined with it in certain proportions: the solution of these mineral substances he conceives to have been promoted by heat, on the abstraction of which they assumed their present solid form. Agricola farther thinks that the precious metals are the results of a more pure and perfect solution *.

Agricola

" tetigeritque metallum, quoddamodo concocta ut item supra dixi. Sed ex luto tenaci potissimum fit lapis."

Lastly, cap. ix. " Igitur proximæ, (sc. causæ), sunt " calor et frigus: deinde quoddam modo succus lapidescens. " Nam lapides, quas aqua humectando dissolvit, eos calo- " rem exsiccando compegisse intelligemus: contra vero, " qui calore ignis liquescunt, ut silices, eos frigore concre- " visse.

* He explains this formation of metals in the following passage of lib. v. cap. vii. " In materia ipsorum (sc. me- " tallorum) inesse aquam, argumentum est maximum, quod " ignis calore soluta fluunt; frigore aëris vel aquæ rursus " densentur. Hoc tamen sic accipiendum, quod aquæ plus " in ipsis sit terræ minus: simplex enim aqua non est " eorum materia, sed mista cum terra. Atque terræ qui- " dem portio tanta in mistura est, quanta aquæ perspicuita- " tem obſcuret, fulgorem non auferat. Quinetiam mistura " quo purior fuit, eo preciosius ex ea fit metallum, magis- " que tolerant ignium. Sed quota terræ portio in quoque

" humore

Agricola is thus the first who has written any thing certain on the formation of veins and the substances contained in them : but, considering the state of chemistry and philosophy at that time, it is by no means surprising that he has sometimes wandered from the right path. Nevertheless his theory of the manner of the filling up of veins has been retained, and is still with some modification very generally received, as will be more particularly seen in the sequel of this chapter. This author rejected the more commonly received opinion of his time, an opinion engendered by the dreams of astrologers, viz. that the planets had an influence in forming the various metals. The opinion which supposes veins, such as we now see them, to have been formed at the same time with our globe, he equally regards as at variance with fact : and he calls it the opinion of the vulgar.

§ 8.

What has been written by Utman, Elterlein, Meier, Löhneis, Barba and others, posterior to Agricola, I pass over in silence, as of too little consequence to deserve

---

"humore ex quo efficitur metallum, insit, nemo mortalium "unquam mente cernere potest, nedum explicare."

deserve any notice; the richness and poverty of veins being by them attributed partly to the position of the mountains with respect to the sun, and partly to the influence of the stars.

§ 9.

After Agricola, the first that deserves any notice is Balthasar Rösler. He seems to consider veins as open fissures which were afterwards filled up *. In his *Speculum Metallurgiæ*, at the beginning of the second chapter, when treating of veins and fissures, he says, "A fissure is a chink or gap which cuts and "divides the rock, and is like the crack in a vessel "which allows water to flow through it; some of "these fissures are long and wide, others short and "narrow." "A vein cuts and divides the rock uni-"formly in a certain direction and inclination, but "its

* Balthasar Rösler's Speculum Metallurgiæ Politissimum; folio. Dresden, 1700.

Rösler was at first *markscheider* and *gegenschreiber* at Freyberg, and was afterwards mine-master at Altenberg, where he died in 1673. He wrote in the last years of his life the instructive work just quoted; but which was not published till 21 years after his death.

" its breadth varies ;" farther on, " what constitutes " the vein itself, and what it contains enclosed with- " in it, and forms its principal part, is either the " materials in which the mineral is found, or metal- " lic matter, or a kind of clay, quartz, spar, &c." " The rock sometimes contains druses ; these are " hollow cavities of a round or oblong form, and " different sizes, which are commonly found in " veins Sometimes the veins are very full of these " cavities, when they are said to be open ; a vein is " said to be shut when it is quite filled up either " with stony or metallic matter. The druses are " often filled up with clay or other matter ; in that " case the vein is said to be closed, although it con- " tains druses."

From these quotations it is obvious that Rösler considers veins and fissures as having the same origin : the latter are open empty spaces ; the former similar spaces, entirely or nearly filled up. He does not explain himself farther, either with respect to the manner in which the fissures were produced, or the mode in which they have been filled up.

§ 10.

Becher, in his Physica Subterranea *, ascribes the formation of metals and minerals to certain subterraneous vapours, which arising from the bowels of the earth, and penetrating the veins, produce a peculiar change in such stony and earthy matters as they meet in their way, which are in a state to be transmuted. He regards the earth as a hollow body, filled with clay, water, sulphureous and bituminous substances; and from this immense reservoir there arise certain exhalations which form the metals. I find, in the writings of Becher, no further account of the origin of the fissures which the veins occupy, nor any observations on the substances of which they are composed.

§ 11.

Stahl, a great physician and chemist of Germany, and commentator on Becher, in many places of his work,

* Jo. Joachim. Becheri Physica Subterranea, Editio novissima, cum præfatione præmissa, indice adornato et specimine Becheriano subjuncto à G. E. Stahlio. Lipsiæ 1703. The first edition of this book was published at Frankfort on the Main in 1669.

work, notices the formation of veins and minerals. In his Specimen Becherianum *, he says, one of the most probable opinions is, that, from the period of the first existence of our globe, considerable fissures had been formed; that these, at the time of the deluge, were filled up with matter of a soft or clayey consistence, which was afterwards penetrated by vapours, extricated from the interior parts of the earth, possessing the property of converting these into mineral matter. This opinion (towards which Becher leaned), he admits to be attended with many difficulties when it is to be applied to particular instances; and that it is often contradicted by what is observed in nature; so much so indeed, that he concludes by rejecting the theory, and considers veins, as well as the substances of which they are composed, as having been formed at the same time with the earth itself; and of course as being of the same age with the rocks in which they occur. At the same time, he is disposed to attribute some effect to the action of

* Georg. Ern. Stahlii Specimen Becherianum, 8vo. Lipsiæ 1703. This was published along with Stahl's edition of Becher's Physica Subterranea.

of the air, and other causes, in producing some changes.

Stahl had already advanced this opinion, in his Treatise de Ortu Venarum Metalliferarum; but he brought it again forward in the work we have just quoted *.

§ 12.

Henkel † is the first who attributes the formation of the contents of veins to a peculiar exhalation produced and engendered by a fermentation, supposed by him to take place in the interior of rocks. The basis of each metal and mineral he supposes to exist already

* Ejusdem Propemticon inaugurale de Ortu Venarum Metalliferarum, 4to. Hal. Magdeb. 1700.

† John Frederick Henkel was born at Merseburg in the year 1679. He at first practised medicine at Freyberg, and became counsellor of mines to the elector of Saxony, then king of Poland. He taught chemistry and mineralogy with uncommon reputation, and died in 1744. He is the father of mineralogical chemistry. His lectures were resorted to by students from all nations, particularly from Sweden. His numerous writings have been read with general approbation, and much used, though less quoted than they deserve. He studied much, and was an unremitting and acute observer. But his style is bad, very diffuse, and tedious.

already in the substance of the rock, and that by a peculiar process of nature it is matured and converted into the metal. He does not venture to ascertain the nature of these bases: but in one passage of his works he speaks of subtile earths; in another of mercurial, arsenical and sulphureous parts. These three last he probably, however, regarded as constituent parts, and consequently metals are compounds. Air, water and fire, appear to him the substances of which nature avails herself in the formation of metals. He also supposes certain kinds of earths and stones to exist, which serve as the matrix for the others, and which are indispensably necessary in the formation of minerals.

This theory, which has been adopted by many mineralogists, is given at large in his Pyritologia *, and in his treatise De Appropriatione †.

§ 13.

---

* Joh. Fr. Henkel's Pyritologia, or History of Pyrites, published at Leipsig, 1725. A second edition of this book appeared 1754. The author treats of the engendering of ores in the 13th chapter, which is written on the original parts of pyrites. The principal passages are § 731, 737, 738, 742, 744—747.

† Medicorum chymicorum non ultimum conjunctionis primum appropriatio etc. invenit et exposuit D. J. Fr. Henkel,

§ 13.

Hoffman, in his Dissertation de Matricibus Metallorum *, (a work written with care and full of accurate observations, but much less read than it deserves), supposes veins to have been formed in the fissures of rocks; yet he only speaks of it as an hypothesis, and in the following words. "Ponamus "itaque venam esse non nisi lapidum fissuram, po-"namus eandem esse cavernosam, certè ubi metal-"lum intra venam produci debet, cortices profecto "erunt

---

Henkel, 8vo. Dresdæ et Lipsiæ 1727. A German translation of this work is found in Henkel's Mineralogical and Chemical Essays, with observations on them by Carl. Fred. Zimmerman. Dresden and Leipsig 1744.

* Jo. Georg. Hoffmanni Dissertatio de Matricibus Metallorum, 4to. Lipsiæ 1738, pp. 55. This excellent work, which contains many good observations, is much less known than it deserves to be.

The author of this little work was born at Leipsig in 1741. He was appointed assessor to the council of mines at Freyberg; and after Henkel's death delivered lectures on chemistry and metallurgy. In 1746 he obtained the rank of commissioner of mines, soon after which he took charge of the mines of the king of Naples, in whose service he died.

" erunt Matrices Metallorum. Jam vero ante om-
" nis Metalli generationem, cunctæ probabiliter erant
" hiantes venæ, suo saltem cortice instructæ, mini-
" mum hinc et inde varii generis saxo et interve-
" niente fissura, occupatæ: ergo cortices venarum,
" aut intra venas existentes lapides, propriè loquen-
" do, universales Metallorum erunt matrices."

§ 14.

Zimmerman, in the second part of his work, entitled " Obersächsische Bergakademie *," is the first mineralogist who considers veins, and the minerals of which they are composed, to have been produced by a transformation of the substance of the rock; and expresses himself in the following manner. " Minerals are undoubtedly formed in the
" rock; but daily experience shows, that the rock is
" not of itself capable of forming a metal: for were
" the

---

* Carl. Fred. Zimmerman Obersachsische Bergakademie, 4to. Dresden and Leipsig, 1749.

Zimmerman was born at Dresden, where he spent his life. He was the pupil of Henkel, whose mineralogical essays he published. Mineralogy and chemistry were his principal studies. A short time before his death, which happened in 1747, he was appointed chief commissioner of mines.

"the mineralising principle capable of converting it "into a mineral, we should find whole mountains "which had undergone this change. But this change "we only meet with in some parts, which follow "certain directions, and being thus transformed, "constitute veins. These veins, when they have "not suffered the entire change, or when they do "not contain perfect minerals, are still of a different "nature from the rest of the rock. An attentive "examination will show, that they are of a decom-"posed and friable nature, appearing to have a ten-"dency to return to this natural earthy state: from "which we may conclude, that these veins were in "reality originally the same as the rock; but that "their texture had been altered and decomposed by "some particular saline substance, which, penetrat-"ing the rents and fissures, had rendered them fit "to be transformed into minerals. In this way "saline substances prepare and render the earthy "matters capable of being converted into the ma-"trix of minerals. These same saline substances "also assist in the formation of metals: to this we "cannot withhold our assent, if we admit, what "experience teaches, that all minerals are mixed "bodies, consisting of a metal, an earth, and an "acid."

§ 15.

§ 15.

It is certainly, in the writings of Von Oppel, formerly captain-general of the mines of Saxony, that the most interesting particulars hitherto noticed on the subject of veins are to be found. He admits positively, and without restriction, the proposition, that *veins were formerly fissures open in their superior part.* In the introduction to his Subterranean Geometry *, (Dresden 1749), he expresses himself on veins as follows.

" DEFINITION.

" § 538. By a *rent* is understood an empty fissure " or gap in a rock."

" REMARK.

" § 539. Rents are in general very narrow, which " is the reason why they are represented to be cracks " or fissures in a rock."

" DEFINITION.

" § 540. A *vein*, on the contrary, is a fissure " which extends a great way through a rock, which

* Anleitung zur Markscheidekunst nach ihren anfangsgründen und ausübung kurzlich entworfen, 4to. Dresden 1749.

" it *cuts* and *divides;* it is filled with materials dif-
" ferent from the rock."

" NOTE.

" § 541. *Fissures* and *veins* do not run in the *di-*
" *rection* of the *strata* of the rock: they traverse and intersect them."

" DEFINITION.

" § 542. The mineral matter which fills the fis-
" sure of a rock is always of a different nature from
" that of the rock; it is called the mass of the vein,
" (Gangart.)"

This same writer is the first who, in the work just noticed, has established the difference between *beds* and *veins*. He settles the distinguishing characters of veins in the § 541. just quoted, where he says that veins traverse and intersect the beds of mountains. It is also of consequence to observe what he farther says on the subject of beds, in the three following paragraphs.

" DEFINITION.

" § 531. A *bed* is a mineral deposit, the na-
" ture of which differs, in part at least, from that
" of the other beds and layers of the same moun-
" tain."

" NOTE.

" NOTE.

" § 533. A *bed* has the same direction as the " strata of the mountain."

" REMARK.

" § 535. I can perceive no impropriety in giving " the name of *beds* to mineral depositions which oc- " cur in mountains whose strata are vertical, or " nearly so, provided they possess the character just " mentioned, viz. that they are of different sub- " stance from the other beds, and have the same in- " clination and position."

I must still quote another passage from the important work of this author, whose memory will ever be held dear, not only by the miners of Saxony, but by every well instructed miner, and will be no less an object of veneration, than his writings will be an everlasting monument of the clearness and sagacity of his understanding and depth of his knowledge. It is a passage, in which he speaks of the utility of researches into the nature and formation of veins; it is valuable in the history of this theory; and it shows what importance those who were considered as skilful miners, and zealous for the advancement of the working of mines, attached to the study of the theory of veins, and how anxious they were to throw some light on a subject of so complicated a nature,

yet

yet so intimately connected with the practice of mining. "The investigation," says he, "of the causes "which have produced the rents in the bowels of "the earth, and of the manner in which they have "been afterwards filled up, are not objects of mere "curiosity, but are highly worthy of the attention of "the naturalist. The study of these objects is indis-"putably the principal, as it is the most useful and "most advantageous one, not only in the science of "veins and fissures, but even of all mineralogy. It "is perhaps the one that has hitherto had the least "attention bestowed upon it."

§ 16.

Before proceeding to notice what Von Oppel has said in a second work on the same principles, though in a more particular manner confined to veins, I shall mention the opinion of some mineralogists who have written on their nature, before his time. The first is Lehman, in a treatise on the Matrices of Metals *. This book is little more than what Hoffman had done on the same subject, as expressed in his § 13th: he has

* D. Joh. Gottl. Lehman's Abhandlung von den Metalmüttern und der Erzeugung der Metalle, 8vo. Berlin, 1753.

has only repeated what had been written by that author on the nature and formation of veins; to which he has added his own opinion, remarkable only on account of its singularity. He begins by saying, (p. 178.) "What is called a rent, is an open fissure "in a mountain, which has been produced by a division of the rocks; and veins are, in my opinion, "nothing but fissures which have been filled by nature with stones, minerals, metals, and clay, in "short, which are of a very different nature from the "rock itself." Farther on he says, "The veins which "we find in mines, appear to be only the branches "and shoots of an immense trunk, which is placed "at a prodigious depth in the bowels of the earth; "but in consequence of its great depth, we have not "yet been able to reach the trunk. The large veins "are its principal branches, and the slender ones its "inferior twigs. What I have said will not appear "incredible, when we consider, that the bowels of "the earth, according to every observation, is the "workhouse where nature carries on the manufacture of the metals; that from time immemorial, "she has been working at, and elaborating their primitive particles; that these particles issue forth, in "the form of vapours and exhalations, to the very "surface of the globe, through the rents, in the

same

"same manner as the sap rises and circulates through "vegetables, by means of the vessels and fibres of "which they are composed."

This explanation of the origin and formation of veins has long since sunk into neglect; and has, so far as I know, met with no followers among writers.

§ 17.

We meet with much interesting matter in the Elementa Metallurgiæ Chemicæ of Wallerius *; although the manner in which he, as well as another Swedish mineralogist, express themselves on this subject, shows that the Swedish mineralogists and miners have very imperfect ideas of the nature of true veins. Wallerius says, (at § 6. of his chap. 3. De Venis Metallicis) "venæ metallicæ extensæ sunt "tractus subterranei in montibus inclusi, ad diversas plagas et diversam profunditatem, recta vel obliqua aut flexa via extensi, in quibus mineræ metallicæ generatæ reperiuntur.

"Obs. 1.

---

* Joh. Gollsch. Wallerii Elementa Metallurgiæ ſpeciatim Chemicæ, 8vo. Holmiæ, 1768.

"Obs. 1. Interstitia quæ intra parietes montis "fissi aut fracti reperiuntur, et à montis fissura vel "fractura dependent, fissuræ appellantur, (Klyfft). "Si hujus modi fissuræ terra, quædam, vel lapide "aut minera, fuerint impletæ, tum generaliter etiam "illæ appellantur venæ. Hoc latiori sensu, vena "est materia fossilis in montibus obvia, in longum "extensa, à circumambiente lapide et saxo diversa. "In sequentibus vero demonstrabimus, saltem non "omnes venas ut à montibus fissura et fractura de-"pendentes esse considerandas, adeoque inter fissu-"ras et venas dari omnino aliquam differentiam." Afterwards in § 9.

"Obs. 2. Vocabulum Schiöl diversimode quoque "usurpatur; alii venam fimbriis donatam appellunt "Schiöl, alii ipsas fimbrias, seu lapidem mineræ ac "petræ interjacentem. Forsan vena, quæ à fissura "originem habet, appellari posset Schiöl; venæ vero "reliquiæ, venæ *."

§ 18.

"Sir Tobern Bergman † gives, in his Physical "Geography,

* This *forsan posset* (perhaps might) shows the uncertainty of Wallerius in this opinion.

† The first edition of the original of this work I have not

Geography, the following short definition of veins. "Miners (says he) apply the name of vein to rents "that are filled up, and consequently are shut; the "words *kluft* and *schiœl* express nearly the same "thing; but what is called *trum*, is a small rent "quite shut, whose walls converging form a kind of "wedge."

§ 19.

I come now to consider the new notions and definitions of veins given by Mr Von Oppel in his Essay on the working of Mines, in which he treats this subject in a more particular manner than in his Introduction to Subterranean Geometry. What he has written

---

seen, nor do I know the year in which it first made its appearance. The German translation was published in 1769. The second edition bears the following title: Physick Beskrifning öfever Iordklotet, förfatad af Tobern Bergman, Andra Uplagan, 8vo. Upsala 1773.

Bergman's Physicalische Beschreibung der Erdkugel (Physical Geography) was published at the desire of the Cosmographical Society; and a third edition, in German, was publiſhed in 1791, at Greifswald. This interesting work contains all that was known on the subject of Physical geography in his time. A work on the same plan, and brought up to the present time, would be of infinite value. T.

written on this subject evinces him to be a skilful naturalist, a profound philosopher, and an experienced miner and geognost. As his remarks on this subject are very interesting, I shall give them at full length.

" § 29. The natural structure of the globe seems " to show us that after the formation of the primi- " tive and principal secondary mountains, they had " suffered great desiccation, and been exposed to vio- " lent shocks. In consequence of these changes, the " rocks and mountains which formerly composed one " continuous mass, were split asunder; whilst this " took place, it might easily happen that one of the " rocks slipt from the other without ceasing to touch " it, or these parts might be separated from each " other, leaving between them open spaces, which " were afterwards filled up, in part at least, with " different mineral substances. The greater part of " these grand events belong to that part of subter- " ranean natural history, which can only be eluci- " dated by a consideration of the facts which the " earth presents to our view; for all these great re- " volutions took place at a period long before the " globe became habitable to the human species. But " whether fissures and veins were actually formed " in the manner we have described, or not; it is no

 " less

" less true that this manner of representing their " mode of formation and the relative situation which " they bear to one another in the mountain, is the " most simple way of accounting for them. It ex- " plains the uniform law of their formation both in " a general and more particular manner, and conse- " quently we shall admit it as the real one. This " hypothesis would be still more satisfactory to the " naturalist, if it were equally easy for him to con- " ceive, how a new mineral substance could be " formed in these fissures, of a nature different " from the rocks in which the veins occur.

" § 30. A *rent* or interruption of the continuity " of a rock, when it intersects the strata, is named a " *fissure*, (gangkluft).

" § 31. A vein is a rent in a rock of which the " direction and position may differ, more or less, " from that of the beds of the rock in which it oc- " curs; and which has been afterwards filled with " mineral substances of a different nature from " those of which the mountain is formed." Further,

" § 38. It is difficult to conceive how rents and " openings of considerable size can be produced " without some portion of the adjoining rock suffer- " ing such a strain as would form lateral chinks, so

" that

" that the principal rent should terminate in several " smaller ones. When these smaller rents are filled " with the same matter as the principal vein, they " are said to be branches of the principal vein " (Trümmer); in this case the vein is said to ramify " or divide into branches. The portion of rock con- " tained between these branches, is for the most " part of a wedge shape; in such cases the includ- " ed mass is called a wedge. When the branches " run for any length in the same direction with " the main trunk, they are called *accompanying* " *branches*.

" § 42. It sometimes happens, that when a vein " is formed in a stratified mountain, the vein not " only traverses, but also deranges a stratum, that is " to say, that one of the two parts of the intersected " stratum changes its position, being elevated or de- " pressed in relation to the other strata; in this case " the strata are said to have shifted. The vein " which produces this effect, is called *(wechsel)* " *shifter*."

It would be necessary to translate all that has been written by this author on the subject, were I to state the whole luminous, instructive, and satisfactory view which he has given of it. I shall content myself with quoting two more geognostic observations which

which are extremely interesting and useful in practice; viz.

"§ 98. In the hollows and valleys of mountains "of a moderate height, the veins for the most part "follow the direction of the valleys.

"§ 104. When a vein has been cut or deranged "by a visible rent, it is again met with in following "the direction of this last, on the supposition that "when the parts of a vein are thus cut, they are "mostly separated from each other.

§ 20.

About a year after the appearance of Von Oppel's work, Delius published a short treatise on the origin of rocks and veins *. The principal object of this work, as expressed in the title, is to give an account of the formation of veins. He considers them to be rents, formed by the drying of the rocks, which have been fince filled up. He is of opinion, that the rain having

* Abhandlung von dem Ursprunge der Gebirge und der darinne befindlichen Erzadern, oder der sogenanten Gänge und Klüfte; ingleichen von der Vererzung der Metalle, und insonderheit des Goldes, 8vo. Leipsig. 1770. Professor Schreiber is the editor of this small treatise.

having penetrated the substance of the rock, has dissolved, suspended, and afterwards carried into the rents, the different materials which serve as a base for stones and metals. The air and heat of the sun, he imagines, are afterwards of use in evaporating the water, and in assisting to combine the materials from which they are produced. According to this author, different metals are formed, in proportion as the original particles have been in a state of greater or less purity *. This subject is discussed by him in a very particular manner, which is any thing but founded on the principles of sound philosophy and chemistry, or supported by correct and satisfactory observations. Besides, it is evident, that all he has said on the formation of veins has been borrowed from Agricola. From an author of the reputation of Delius, who was acquainted with the writings of Von Oppel on the same subject, we should have expected to meet with observations more worthy of notice.

In

* This author treats of the subject of veins in the second and third sections of the above work; the second is written on the origin of veins and rents; and the third, on the probable mode of formation of the ores that are found in them, from § 33. to 86. The principles of this theory are contained in from § 35. to 37. and 61. 64.

In his treatise on the working of mines, which appeared some time after his other work, the same author treats of veins in a still more circumstantial manner, but with as little solidity, and upon the same principles *.

§ 21.

The theory of the formation of veins, which the sub-director of mines, Mr Von Charpentier has delivered, at the end of the fourth section of his Mineralogical Geography of Saxony †, is nearly the same with that which has already been considered in § 14. The

* Christoph. Traugott Delius Anleitung zur der Bergbaukunst nach ihrer Theorie und Ausübung, nebst einer Abhandlung von den Gundsätzen der Berg-und Kameral-Wissenschafft, 4to. Wien. 1773.

The subject of veins is treated of from § 13. to 52. of this work.

In this work, the author shows in several places some disrespect for Oppels Bericht vom Bergbau.

† Joh. Fr. Wilh. Charpentier's Mineralogische Geographie der Kursachsischen Lande, 4to. Leipsig, 1778.

This is unquestionably the best of Charpentier's writings; and although published so early as 1778, is an excellent specimen of topographical mineralogy, and ought to be known and studied by every mineralogist. The style of writing is good, the arrangement judicious, and the descriptions often excellent. T.

The concise manner in which he has delivered his opinion on this subject, prevents us from giving an extract from it, and to insert the whole would greatly exceed the limits which have been prescribed to this treatise. I think it also unnecessary to enter into any particular detail of it, as the excellent work in which this theory is delivered, I suppose to be in the hands of all my readers. The explanation of it will be found from the 425th, to the 432d page of the work in question.

I may however remark, that in the section immediately preceding that in which he begins to explain his theory, Mr Von Charpentier has brought under one view, all the strongest objections that can be urged against the theory which confiders veins to have been rents that were afterwards filled up by different mineral substances.

§ 22.

Among the different authors who have written on the theory of veins, Baumer, counsellor of mines, also merits a place. He treats of this subject in his Geographia et Hydrographia Subterranea *. What he

* Fundamenta Geographiæ et Hydrographiæ Subterraneæ, à Jo. Guil. Baumer, 8vo. Giessæ 1779.

he says on this subject, though short, comes very near the truth, as the following quotation from chap. xiv. § 14. will show. "Veins differ from the strata "of the rocks in which they occur, both with re-"spect to their form and their substance. Their "formation is posterior to the rock. It appears, "from many observations, that they have been "formed under the ancient sea; for their upper ex-"tremity is often covered with several beds of schis-"tus, and in cavities in the substance of the vein we "frequently meet with marine animals in a petrified "state *."

§ 23.

Charpentier's particular theory is further illustrated in his Beobachlungen über die Lagersládle der Erze. Leipsig 1799. T.

* The following is the whole passage. "Cum à mon-"tium stratis venæ tam materia, quam forma, differant; "hos alio tempore et serius, quam ipsos montes structas "esse, arbitror. Hanc vero structuram jam in mari veteri "absolutam fuisse, pluribus argumentis demonstrari potest: "1. Venarum ora sæpius multis strata contecta sunt, id "quod in Hassia nostra admodum promiscuum est, et venas "inventu difficiles reddit. 2. Tam in cryptis vacuis, quam "venis, petrefacta marina quandoque inveniuntur. 3. In sic-"ca telluris superficie nulla crypta vacua mineres repletur." And in the end of the same section, he says. "Terræ ſubti-

"les

§ 23.

Mr Gerhard, privy counsellor of finance, in his Essay on the History of the Mineral Kingdom *, treats of veins and their mode of formation in a very particular manner, and has collected a number of highly interesting and instructive facts on this subject. This author, like most other geognosts, considers veins (which he distinguishes sufficiently well from beds,) to have originally been rents in mountains, which were afterwards filled up with mineral substances. According to him, many causes may have contributed in forming these rents, which he

F supposes

" les &c. aqua marina stratorum fissuras cryptis advectæ,
" venis mateiiam præbuisse videntur."

The observation of the occurrence of petrifactions in veins is a very important one, and would have been of still greater consequence had the place been mentioned where they do occur.

* Carl. Abraham Gerhards Versuch einer geschichte des Mineral-Reichs. Ersther Theil, 8vo. Berlin, 1781.

All that relates to this subject will be found in the fourth chapter of this part, from § 117. to 148. The whole of the chapter is occupied with the place and manner in which minerals are found in the earth, more paiticularly in mountains.

supposes to have been produced at very different periods of time. He is disposed to think that subterranean fermentations may have contributed towards producing rents, which having been afterwards filled, became veins. In order to explain the manner in which veins have been formed, he has recourse to the supposition of water penetrating into the substance of the adjoining rock, dissolving certain particles with which it becomes loaded, and then, after passing through the crevices of the rock, depositing these particles in the rents which we now find occupied by veins. Minerals he supposes to have originally existed in the rock, and that they had been carried in a fluid state into those places where we now find them.

§ 24.

The sub-director of mines, Mr Von Trebra, in that instructive and splendid work, entitled Observations on the Interior of Mountains *, gives many important

* Erfahrungen vom Innern der Gebirge nach Beobachtungen gesammelt und herausgegeben von Fredrich Wilhelm Heinrich von Trebra mit vielen illum Kupf. fol. Dessau und Leipsig, 1785.

tant observations on veins, accompanied with a theory of their formation very much resembling that of Zimmerman already noticed at § 14; he treats of this subject in the beginning of his work, in a series of letters addressed to Mr Von Veltheim, director of mines. The whole theory of the formation of veins and minerals forms the subject of the third letter, in which he treats of the circulation of fluids in the following words.

" In explaining the phenomena which are obser-
" vable in the interior of mountains, (it must how-
" ever be remembered that I do not include such as
" are evidently of volcanic origin), I do not avail
" myself of those great causes, which by their mag-
" nitude, the suddenness of their action, and by
" their effects, produce sudden changes which take
" place under our eyes, such as subterranean fires,
" earthquakes, and the like. I refer these phenomena
" to natural causes, which, though less evident and
" slower in their operation, are no less certain of
" producing a radical transformation. Of this kind
" are *putrefaction* and *fermentation*. It is of little
" consequence by what name we distinguish this
" peculiar action exerted by nature in the mineral
" kingdom; it consists in an intestine motion in the
" central parts of the globe, and appears to be pro-

" duced

"duced by water combined with heat in different "degrees of intensity. I observe such changes still "going on, and can conceive them to continue so "long as the same series of operations exists in na- "ture. I am persuaded that there is constantly go- "ing on in our mountains a variety of transforma- "tions, compositions, and decompositions, which "not only take place at present, but will continue "to the end of time."

The same author, a little farther on, says, "Fer- "mentation, if I may be allowed to call by that "name, this quality which acts by insensible de- "grees, produces the most perfect transformations "in the bowels of the earth; fermentation, I say, "may, according to my theory, alter the entire mass "of a mountain; it may convert *granite* into *gneiss*, "as this last only differs from the former in its "structure, which is slaty or schistose; gneiss, in- "deed, has no other distinctive character than its "structure, namely the regularity and parallelism of "of its beds, and in some places a decomposed feld- "spar approaching to clay. This fermentation may "also convert *grey wacke* into an argillaceous schist, "which last may again, by induration, become jas- "per, when this process is either diminished or "stopt. By it also quartz may be converted into "clay,

" clay, calcareous substances into quartz, and the " whole mass of a mountain into inflammable or saline matter, or even into ores, metals, or semi-metals. To it I ascribe the power of producing, preserving, and continuing to form, the different beds " and mineral repositories which are found both in " primitive and floetz mountains: finally the effects which the waters produce in filtering from " above to below, and which in their passage through " the different rocks may undergo some peculiar modification, appear to me the principal cause why " this fermentation may act with more force in " one part of the same mountain than in another.

" I shall now proceed to give the definition of " *beds* and *mineral repositories*. They are certain " parts of a mountain, where, in consequence of an " intestine motion occasioned by the afflux of the " water, the rock, and the different foreign bodies, " whether of the animal or vegetable kingdom, which " they contain, are converted into a stoney or metallic matter which is now no longer the substance " of the rock."

§ 25.

§ 25.

The last author I am acquainted with who has given a particular theory on the formation of veins, is Lieutenant Lasius in his observations on the mountains of the Hartz *. He considers veins to have been formed in the rents which had been produced by the revolutions and other effects of nature. These rents he supposes to have been afterwards filled with water, which, by containing carbonic acid and other solvents, acquired the property of dissolving such earthy, metallic or other matters, as they encountered in their percolation through the different strata of the rocks which they passed through; these were afterwards precipitated by certain substances which had this property, and thus were deposited in the particular cavities where they are now found. But this author is uncertain upon one point, namely, whether the waters found the metallic particles already formed in the substance of the rock, or whether they engendered

* George Otto Sigismund Lasius Beobachtungen über die Harzgebirge, als ein Beitrag zur Mineralogischen Naturkunde, 11.—12, Theil, 8vo. Hanover, 1787.

This theory of veins is in two parts, and in this edition is contained in from p. 413. to 427. The principal part is in p. 415. to 418.

gendered them; for after having delivered the former of these opinions, (he adds), "or whether the "solvent produced on the minute metallic seeds (if "I may so express myself) certain modifications "and changes by which they were in one place "converted into *lead*, in another into *silver*, or any "other of the metals or semi-metals."

§ 26.

I cannot conclude the history of the different theories which have been proposed to account for the formation of veins, without remarking that it is to the Saxon philosophers and miners that we are indebted for these theories. Agricola, Rösler, Henkel, Hoffman, Oppel, Charpentier and Trebra have furnished all the information we possess on this subject. Nor is this surprising, when we reflect, that it is chiefly the philosophers and miners of Saxony who have formed a distinct branch of science of the art of mining; who have carried their researches on this subject the greatest length; and who have formed some of the accessory sciences, as the art of smelting ores, and subterranean geometry *.

§ 27.

* The art of assaying has been much improved by the labours

§ 27.

Having in this chapter shown the variety of theories which have been proposed to account for the formation of veins; I shall in the third explain my own theory: the proofs on which it rests will be given in the fourth, fifth and sixth chapters; and in the eighth I shall state the objections which I have to urge against the theories which were given prior to it.

---

labours of George Agricola, Lazarus Erkern, and Modestin Feachsen. The two last, in particular, have written very useful works on this subject. In modern times, the counsellor of mines, Gellert, has written a very useful treatise on the same subject. Other countries have also produced similar works; Cramer in Lower Saxony, and Scheffer and Bergman, two Swedes, deserve to be mentioned. The latest work of this kind is that of Aulic-counsellor Gmelin.

The science of subterranean geometry was formed by *Agricola*, *Rheinhold*, *Rösler*, *Weidler*, *Voigtel*, and *Baier*, successive; but brought to perfection by *Oppel Kästner*, and *Scheidhauer*. The most recent work on Subterranean Geometry is that of *Professor Lempe.*

CHAP.

CHAP. III.

SHORT ACCOUNT OF THE NEW THEORY OF VEINS, AND OF THE MODE OF THEIR FORMATION.

§ 28.

ALL *true veins* were originally, and of necessity, rents open in their upper part, which have been afterwards *filled up from above.*

§ 29.

Rents may be produced by many different causes. Mountains have been formed by a successive accumulation of different beds or layers placed or heaped upon one another. The mass of these beds was at first wet, and possessed little solidity, so that when the accumulation of matter had attained a certain height, the mass of the mountain yielded to its weight, and must consequently have sunk and cracked. As the waters which formerly assisted in supporting the mass of the mountain began to lower their level; these masses then lost their former support, yielded to the action of their weight, and began to separate and be detached from the rest of the mountain, falling to the free side, or that where the least resistance was opposed. The shrinking of the mass of a mountain, produced by desiccation, and still more by earthquakes, and other similar causes, may also have contributed to the formation of rents.

§ 30.

The same *precipitation*, which in the humid way formed the *strata* and *beds* of rocks, (also the minerals contained in these rocks), furnished and produced the *substance of veins*; this took place during the time, when the solution from which the precipitate was formed, covered the already existing rents, and which were as yet wholely or in part empty, and open in their upper part.

§ 31.

Veins (whether considered as rents, or as the substance constituting the vein,) have been produced at very different times, and the antiquity or *relative age* of each can be easily assigned.

The

The *distinguishing characteristics* for the *relative age* of veins, and their substances, are the following.

1. Every vein which *intersects* another, is *newer* than the one traversed, and is of *later formation* than all those which *it traverses*; of course the *oldest vein is traversed* by all those that are of a *posterior formation*, and the *newer* veins always cross those that are *older*.

When two veins cross, one of them without suffering any derangement or interruption traverses the other; this last is interrupted and cut across through its whole thickness by the former. The first of these is said to traverse the other, and the latter to be traversed by the former. The vein which crosses another is of newer, whilst this last is of older formation. See § 35. and 45. This crossing of veins is of great importance, and deserves to be kept in remembrance by all who wish to become acquainted with the study of veins; yet, till now, it has always escaped the observation of mineralogists.

2. The *middle part of veins* is commonly of later formation than that portion which is nearest their walls; and what we find in the *upper part* of a vein is *newer* than what we meet with in the lower part.

3. In a specimen composed of different minerals, the

the *super-imposed portion* is always of newer formation than that on which it rests, which is of course older.

In compound masses, more particularly when composed of different cryſtallizations, it is always of consequence to observe the position and arrangement of the different minerals, that we may be able to ascertain the relative age of each. These observations are particularly necessary in determining the formation of minerals in general. I have already laid great stress on this in some of my other writings, as in giving the external characters of *apatite* * and *olivine* †.

§ 32.

It is very easy to ascertain and distinguish the different *vein-formations*, taken individually, in whatever place we meet them.

When *veins*, even in distant tracts of country, contain the *same* ores and vein-stones, and when these are

* See the Bergmanisches Journals, Part I. first year, (Freyberg 1788,) for the history, character, and chemical analysis of apatite, by A. G. Werner, pages 90, 91.

† See Bergmanisches Journal for 1790, Part II; for the external character of olivine and chrysolite, pages 62, 63.

are arranged in the same determinate order, we conclude, that they belong to one and the same general formation. Veins are ascertained to be of the same formation, when the materials of which they are composed are of the same nature. The *greater the variety* of substances contained in the vein, so much the *easier* is the determination of this point.

§ 33.

1. All particular enrichments of veins depend, for the most part, on their being filled with ores or metals, whether this has been effected by

a. A particular filling up from above, or

b. By particular internal canals, or

c. By an infiltration across the mass of the vein.

2. A metallic vein may be increased by the junction of a new metalliferous vein.

3. Sometimes, though rarely, the richness of a vein may be the effect of an elective attraction or affinity of the neighbouring rock.

This last cause seems to have taken place at Kongsberg in Norway; the veins are richer in ores there, when they traverse particular beds of the mountains which differ much from the others, and are in that country called Falbänder.

§ 34.

§ 34.

Veins occur more or less numerous in particular districts, which last differ in extent. They are also only particularly metalliferous in certain places. The occurrence of veins in such countries depends much upon the external form of the mountains.

1. On the position of the whole chain of mountains in respect to its extent and declivity.

2. On the particular position of the country where they occur.

a. Whether the country be composed of hills with gentle declivities, and roundish or flattish summits; or,

b. Whether it be a place in a principal valley.

§ 35.

In one and the same country, we often find a number of veins of very different formations; these constitute what is called a *mining district.*

The veins of different formations which often occur in the same tract of country, do not only present traces of a very different formation, but they often show, in the most distinct manner, the distinguishing characteristics of the age of their formations.

In the 10th chapter of this treatise, I shall give an

account

account of the mining district of Freyberg, in which will be observed two classes of veins very different from one another. One of these classes consists of veins which run from N to S: the veins of this contain lead glance; black blende; iron, copper, and arsenic pyrites; quartz, and brown spar. This formation, which constitutes the first deposition of metallic veins in the district, will be more particularly described in the 10th chapter. The second class of veins which always traverse the former, and are never crossed by them, contains lead glance, radiated pyrites, heavy spar, fluor spar and quartz: they stretch between the sixth and ninth hours of the mining compass. These will be taken notice of in the chapter just mentioned, under the third depot of veins.

The mining district of Ehrenfriedersdorf contains veins of tin and silver glance. The tin veins are always traversed by the silver; the direction of the first is between the sixth and ninth hour, that of the last from the ninth to the third hour.

§ 36.

The substances which constitute the veins of a particular formation sometimes occur in different manners.

1. Not

1. Not only *in the veins* peculiar to this formation, but also,

2. In the *intersection* of two veins of a very different nature; often too,

3. In the *middle*, and sometimes, though seldom, on one of the *walls* of another vein.

§ 37.

All the propositions which have been advanced, will, I trust, be sufficiently explained and established, not only by the proofs to be adduced in the following chapters; but still farther by the application of them in the 10th chapter to the mining district of Freyberg. A farther account of this theory will be given in the fifth and seventh chapters.

CHAP.

CHAP. IV.

PROOFS, THAT THE CAVITIES WHICH ARE NOW OCCUPIED BY VEINS WERE ORIGINALLY OPEN FISSURES.

§ 38.

In defining veins *to be rents produced in rocks, which were afterwards filled from above with certain substances*, two effects of nature very different from one another are admitted, both of which, however, are essentially necessary for the formation of a vein; which two effects have been produced at different periods of time. In the first place, the fissure which veins now occupy must have been formed; and, secondly, this fissure must have been afterwards filled up.

To establish the truth of the hypothesis advanced on the formation of veins, each of these effects, supposed necessary to their formation, must be proved.

It must therefore be shown, that veins, in respect to their size, their position, and their relation to each other, bear a striking resemblance to the fissures of rocks; and, that in the length of time which has elapsed since the formation of our globe, such fissures must have been produced, and that they are daily forming. But still farther, that substances of the same nature with those constituting the mass of veins are met with in other situations; that these substances have been deposited in the places where they are now found, by a precipitation in the humid way; that, at the time of their precipitation, they have been introduced into the fissures which formerly existed, and were fit to receive them; and lastly, that the arrangement and disposition of the different substances forming the mass of the vein, is such as shows that they have been introduced from above.

I begin with the proposition advanced in the preceding chapter, at § 25. and 26.

The cavities which veins now occupy, are rents which have been formed in the rocks; these cavities were at first fissures, gaps, and rents of various size, with openings in their upper parts. In support of this position, I shall bring forward the *nine following proofs*, which, I hope, will remove all doubt of its truth from

from the mind of every intelligent and unprejudiced geognost and miner.

§ 39.

FIRST PROOF.

When the mass of materials of which the rocks were formed by precipitation in the humid way, and which was at first soft and moveable, began to sink and dry, fissures must of necessity have been formed, chiefly in those places *where mountain chains and high land existed.*

As, in the first place, these accumulated materials were not equally dense, nor arranged everywhere of the same degree of thickness and height, it follows of course, that this sinking did not take place in an uniform manner. From this difference in the sinking and shrinking of the solid materials of our globe, different separations, rents, and fissures, have been produced.

Secondly, These fissures and rents must, of course, occur more abundantly, where the greatest quantity of matter has been heaped up, or where the accumulation of it has formed those elevations which are called mountains; because these, being more at liberty, and having less support at their sides, must yield to the consolidating effects of pressure arising from

from their own weight, which must have produced interruptions of the continuity of the mass, in consequence of which veins were formed in the lowest parts; for the mass of the mountain would naturally incline over to that side where it found the least resistance *. In this way we constantly observe similar rents produced, though on a smaller scale, in the drying of substances possessing a greater or less degree of humidity.

The confusion and disorder which we often find to prevail in the strata of mountains, is an evident proof of these vast sinkings, an example of which is to be seen in the strata of a conglomerate which occurs in the coal formation of Hainichen. This conglomerate is almost solely composed of flat pieces of clay slate, and of considerable size; having in some places a position nearly vertical, the same with that of the strata themselves. Now it is impossible that these stones should have been arranged in this manner by the action of the water; they must have therefore taken their position afterwards, along with the strata.

Lastly,

* Compare what is here said with § 51. where mention is made of the veins at Saalfeld.

Lastly, In mountains of newer formation, we frequently meet with rents and fissures still empty, which are even several inches wide. I have not only often met with these myself, but I find the same thing frequently mentioned in the writings of mineralogists. I shall only quote one instance of it from a skillful and cautious observer, Charpentier, (in his Mineralogical Geography, p. 359.), who examined them himself. These are open fissures in the mountains of Kifhäuser, six, eight, or more inches wide: one near the old castle is stated by him to be several feet wide *.

§ 40.

SECOND PROOF.

Rents and fissures are still forming, from time to time, in mountains, which have a close resemblance to those spaces now occupied by veins. This happens chiefly in rainy seasons, and from earthquakes.

In the year 1767, which was a very rainy season, a very remarkable fissure was produced near *Hainichen*, to the eastward of the town. In Upper Lusatia, near the place of my nativity, two remarkable events

* Mineralogical Geography of Saxony, p. 359.

events of a similar kind happened during the same year, both of which I myself have seen. The one happened at *Wehrau*, where part of a mountain composed of sandstone sunk several yards, carrying along with it the largest trees; a narrow fissure was on this occasion formed, above 200 feet long. The other happened two leagues farther on at Tiefenfurth in a sandy soil, where a fissure was formed, a quarter of a league long, three or four inches wide, and so deep that the longest poles could not reach the bottom. A similar thing happened at *Aussig* in Bohemia; this event, to the best of my recollection, was attended with a falling of the mountain. Notice is taken of it in the public prints of the day.

At the time of the great earthquake in Calabria, rents and fissures were formed in great number in that unfortunate country, accompanied with sinking of the mountains. The particulars of these events are narrated in the different works which give an account of that earthquake. We likewise find mention made by different writers of similar rents and falling of mountains which have happened in the Alps, in the Tyrol, in Savoy, and in Switzerland.

It must at the same time be obvious, that the number of fissures now formed, cannot be nearly so great as those which took place at the first formation of of mountains; for they cannot be so easily produced in rocks which have attained a more firm and compact structure than they possessed soon after their first formation.

§ 41.

THIRD PROOF.

Veins, in respect of their form, situation, and position, bear a strong resemblance to rents and fissures which are formed in rocks and in the earth; that is to say, *both have the same tabular figure, and the deviations which they make from their general direction are few in number, and very inconsiderable.*

Veins, like rents, grow narrow towards the lower part, and terminate like a wedge; their lateral extremities terminate in the same manner. Small branches and threads often arise from their walls and extremities. When they are of considerable size, collateral and accompanying veins frequently proceed from their hanging and lying.

With respect to their position, veins are either entirely vertical, or they have an inclination which approaches nearer to the vertical than the horizontal line; most of them have an inclination which corresponds with the declivity of the mountain; finally, all

all the veins of a mining district, more particularly when they are of the same formation, have a similar direction, which shows them to have been produced by the same general cause.

§ 42.

FOURTH PROOF.

No one can doubt that the small oblong cavities (gang-klufte) which are found in great abundance in rocks, are in reality small rents or chinks; now there exists an uninterrupted chain, from the narrowest fissure to the greatest vein, so that it is impossible to draw a line of distinction between what is actually to be considered as a true rent or fissure, and that which is a vein without being a fissure.

We sometimes meet with a small vein, which does not exceed the thickness of a straw, completely filled with mineral matter; and at other times we find rents of three or four inches wide, which are perfectly empty.

§ 43.

FIFTH PROOF.

Are not druses, and the small crystals which line their walls, *only certain parts of a vein which have not yet*

yet been filled up, and consequently the remains of the space in which the vein has been formed?

They have the general direction of the vein; are sometimes several fathoms long, and proportionally wide; are commonly found in the broadest parts of the vein, and evidently appear to have formerly possessed greater length and breadth; but in consequence of a deposition of new matter happening repeatedly on their walls, they have become shorter and narrower, and in some instances almost entirely filled up.

§ 44.

SIXTH PROOF.

A consideration of the materials of which many veins are composed, proves, in so incontestable a manner, that veins were originally empty fissures, that no doubt can now remain on the subject.

The proofs by which this is supported are,

1. That certain veins are filled by rolled masses, or water-borne stones. How could these have gained admission into the internal parts of veins, if we do not allow them to have been originally open in their superior part?

I found a vein of this kind filled with rolled pieces only, in Daniels-stollen at Joachimsthal; it was, in

to work out the vein *Elias*, and carrying on the operation towards the vein *Schweizer*, that this vein consisting of rolled pieces was discovered at the depth of 180 fathoms. This vein was 14 inches thick, accompanied the vein *Elias* for some time, and was almost entirely composed of rolled pieces of gneiss of different sizes, some of which had acquired a shape almost spherical *. I afterwards met with a similar occurrence in the *Stoll-Refier* near *Riegelsdorf* in *Hessia*, where a vein of cobalt, having a direction nearly vertical, was intersected by another vein almost entirely composed of sand and rolled pieces. According to the observations made by Mr Schreiber, veins occur in the mountains of Chalanches, near Allemont in Dauphiné, which are entirely filled with rolled pieces †.

§ 45.

2. That fragments of the adjacent rock are often found

* I have taken notice of veins filled with rolled pieces, in a short treatise, giving an account of the Puzzen Wakke at Joachimsthal, published in the first part of Crell's Chemical Annals for 1789, page 134.

† Bergmanisches Journal, Part I. first year, page 27.

found in the middle of the vein, which shows the repositories to have been open fissures.

These fragments of the rock have actually the form of debris detached from the walls of the vein into which they have fallen. When they are of considerable size, their position is parallel to that of the rock, or to the roof and walls of the vein; which incontrovertibly shows them to have been detached, and, as it were, pushed into their present situation without having been rolled. But when these fragments are of a smaller size, they assume all kinds of directions, which proves that they had been tumbled confusedly into an empty space. It may be remarked, that it is only in rocks possessing a slaty or foliated structure, as gneiss and clay slate, that we find this peculiarity in the arrangement and position of the fragments and debris of rocks found in veins.

At Joachimsthal, I observed two veins filled with fragments, lying confusedly, and without any determinate order. These, as far as I recollect, are the veins called *Geschieber* and *Huber*. The same thing is also to be seen in the Freyberg district, at *Rothenfurth*, in the vein Samuel in the *Isaak Erbstollen*. About 15 years ago, I observed a mass of *gneiss*, about a fathom long, and six inches thick, lying in a transverse direction in the same vein. The singularity of this

appearance

appearance made me point it out to several persons, and I could have wished that it had been preserved; but several years after, this remarkable fragment was removed in working out the vein. In almost every vein we meet with larger or smaller fragments of the adjoining rock. And, if I may judge from such specimens as I have seen in cabinets of minerals, they must occur in great abundance in the lead veins of *Stolberg*, *Strasberg*, in the *Hartz*. In the mine *Gott hülft gewiss*, near *Könitz*, in the county of *Schwarzburg*, a great quantity of the fragments of a schistose rock is found lying confusedly in the middle of copper pyrites.

§ 46.

This originally open state of veins is shown,

3. By the occurrence of the debris of the substance of a vein, which is sometimes found in considerable quantity in the same vein; and which, when mixed with another fossile substance, forms a true breccia (*trümerstein*). These fragments may have been introduced into the vein in one of two different ways. First, The vein, after its first formation, may have been again opened up; that is to say, a new fissure may have been formed in the vein, in the direction of the original one: or, secondly, The first vein

vein may have been intersected by one of newer formation. In these two ways a portion of the vein already formed may have been shattered, and the fragments have fallen into the formed rent; this (which is supposed to have been produced by some violent concussion) must have happened at the time when the solution, from which the substance of the mineral destined to fill the new fissure, was present.

I have frequently met with phenomena of this kind. An attentive observer will often be able to remark the same thing in examining specimens of breccia formed by the matter of veins, which are often to be met with in mineralogical cabinets. I shall give three examples of this kind from specimens in my own collection. The first is the well known famous agate in breccia from *Schlotwiz* near *Kunersdorf*. It is composed of different sized fragments of a striped agate joined together by a cement of amethyst and quartz, and forms a vein of considerable size: in polished specimens small pieces are to be seen, the parts of which correspond so much, that it is evident that they are fragments of the same piece. The second came from the mine *Hulfe-Gottes* at *Memmendorf*, not far from *Oederan*.

A

In one of those veins there is a breccia formed by an assemblage of fragments of different kinds of vein-stones, as heavy spar and radiated pyrites, mixed with brown blende and lead-glance; these substances are united by a kind of quartz of a spongy and friable nature. The workmen look upon this as the old man, (a name given by them to the stones thrown out of an old mine), but the overseer considers them as real minerals; which last opinion is put beyond the possibility of a doubt by the appearance of the substance, as well as by its local position. This rock appears to have occupied a considerable portion of the vein towards its roof. I was much disappointed on my visit to this spot, at being deprived of the satisfaction of examining more minutely into the particulars of this singular occurrence, in consequence of the vein being filled with water, and at being reduced to the necessity of being contented with the inspection of a variety of specimens selected from a heap of rubbish in the vicinity of the mine. The lapidaries have cut and polished for sale many small pieces of it. The third specimen is from the mine *Segen-Gottes* at *Gersdorf*, where it was first observed by Mr Mende, the engineer, who presented to me a specimen of a fragment from the vein. It consists almost entirely of small

pieces

pieces of heavy spar united by a cement of bluish grey-coloured fluor spar, which is in some parts crystallized. I afterwards descended into the mine, and found in the roof of the vein a large opening, a quarter of a fathom wide, and several fathoms in depth and length, filled with fragments of the mass of the vein, which had been afterwards agglutinated and joined together. I had before found breccias of the same kind in an old mine at *Lorenz Gegentrum*, also in several veins at Joachimsthal. Examples of this occur also in some of the mines of the *Hartz*, chiefly in those of *Ring* and *Silberschnur* at Zellerfeld.

§ 47.

4. The occurrence of petrifactions in veins is another proof that they have been originally open fissures. For if we admit petrifactions to be the remains or impressions of organic bodies, (which has never yet been doubted,) these organic bodies must of necessity have either existed in the cavities now occupied by the veins or they must have been introduced into them in their present petrified state; on either of which suppositions it is evident that veins must have been open.

Baron Born, in his letters on different subjects of mineralogy,

mineralogy written during his travels in Hungary, mentions that he found *porpites* at the *Spitlar-Haupt-Gang* *.

Baumer, in the part of his work formerly quoted, mentions that it is not an uncommon thing to find petrifactions in veins. But what I am going farther to quote, comes from a mineralogist who has studied the subject with much attention, and on the accuracy of whose observations the greatest reliance may be placed †. At *Lobberg*, on the *Unstrut*, in floetz limestone

---

* Ign. Von Born's Letters on Mineralogical Subjects, written during his Travels through the Bannat of Temeswar, Upper and Lower Hungary, published by Joh. Jac. Ferber. Leipsig, 1774. He writes on this subject in the following manner. "I must here take notice of one of "the greatest curiosities which I met with, which was a "species of petrified porpites observed by me in the principal spitlar vein, (Haupt-gang,) inclosed in compact cinnabar in a gallery in this vein, at the depth of 89 fathoms, "reckoning from the mouth of the shaft Elizabeth." The miners assured him that they often met with similar occurrences.

The porpites is a kind of madrepore, which is a species of coral.

† This was communicated to me by Mr Von Schlotheim of Niederdorfstat in Thuringia, who observed the fact himself, and examined it with attention. The account which he sent me of it deserves to be given at large; it is as follows.

stone rocks, veins of marl five or six inches wide occur, containing petrifactions of a nature quite different from those which are found in the mass of the rock. In the neighbourhood, beds are said to be observed which are composed of the same marl, and which contain similar petrifactions.

---

lows. " At Lohberg on the Unstrut, between Nagelstadt " and Vargel, in Saxon Thuringia, where the floetz " limestone is exposed to view, the beds of compact " limestone are cut in several places by vertical rents; " these rents are in part open fissures, and in part fill- " ed with friable marl, in which petrifactions occur in " good preservation. It is remarkable that the compact " limestone in the vicinity of these rents (which are from " five to six inches wide,) either shows no traces of petri- " factions, or is full of trochites, whilst the rents contain " cornua ammonis, terebrates and turbinites. On Holz- " berg, situated in the neighbourhood, between Ballstadt " and Beirgtona, we find the same petrifactions in thin " beds of marl, which alternate with thick beds of lime- " stone. But here the limestone contains also cornua am- " monis, terebratulites and turbinites. The rents which are " described as vertical at Lohberg, have a direction quite " different from the inclination of the beds of the mountain, " and are not the same beds which have assumed a verti- " cal position in consequence of the falling of the beds of " limestone. Just at Holzberg, the beds have in most

§ 48.

To add further,

5. *Rock salt and coal*, substances of very recent formation, and which are found in veins, afford another proof that veins have originally been open fissures. How can we account for the presence of these substances (whose origin is well known to us) in veins, if we do not admit them to have been brought from without? From this, veins must have been originally open empty spaces.

Veins of this kind occur very seldom. But I have met with a vein of coal, at *Wehrau*, in Upper

"places an inclination of from 75 to 80 degrees; whilst "here they lie almost horizontal, and it is therefore much "easier to distinguish the rents, or rather the veins. I "must here remark, that these veins descend without interruption to a considerable depth, and only go a small "way into the mountains. The place where they are best "seen, was a few years ago almost entirely carried away "by a great flood; here we can, in many places observe "that the veins go to a considerable depth, but do not "extend into the interior of the mountain, and that they "disappear in proportion as the rock becomes more firm "and compact."

per Lusatia *, a quarter of a fathom thick, which is almost vertical; it occurs in a sandstone rock, in that country called the Devil's Chamber, (Teufels-stuben). This vein contains several inches of very pure coal, but the rest is mixed with much sand. Small veins of rock salt frequently occur in the salt mines at Aehlen in the canton of Berne.

§ 49.

In the last place,

6. The occurrence of the materials of mountain rocks in veins, shows that they were originally open fissures. Of this kind are veins of granite, porphyry, limestone, basalt, wacke, greenstone †, and such like. For

---

* This remarkable coal vein appears at an inaccessible part of the rock, where it is very steep, at the foot of which the rivulet Queis runs. I pointed it out to Mr Von Charpentier, counsellor of mines, who immediately recognised it for a true vein. He has described it in his Mineralogical Geography of Saxony, page 7. In the same rock are to be seen several other smaller veins, not more than an inch wide, which traverse it in a variety of directions.

† I distinguish *greenstone* from *sienite*, and give the former name to those hornblende rocks that occur in the trap-formation, which like basalt, composes the tops of hills in over-lying stratification; it is commonly mixed with feldspar,

For if we admit these substances to have been deposited successively by precipitation in the humid way, and have thus constituted the mass of rocks; we cannot suppose them to have had a different origin when they occur in veins. And, if these substances have been formed by precipitation in the places where they are found, it follows that what we now find as veins were formerly open spaces.

True

---

spar, seldom with mica. The hornblende in this rock generally occurs in small grains, seldom in large grains, often in very small grains. This last variety is commonly more or less mixed with the matter of basalt, so that it passes into true basalt. I have found greenstone rocks in the basaltic mountains at Lobauer in Upper Lusatia, at Weissener in Hesse, and at Dransfeld not far from Gottingen. At Weissner it occurs on the Kolbe (or Kalbe as it is there pronounced), very well marked; it consists of hornblende in large grains, mixed with much feldspar well characterised. It is there called *dukstein*. It occurs in Sweden, in the rocks of the trap formation, (by which name we may distinguish the rocks of basalt, porphyry slate, amygdaloid and greenstone), according to the observations communicated to me by Mr Napion on the mountain *Taberg* in Sweden. An account of these, with observations which I have added at considerable length on the same subject, will be found in Bergmanisches Journal, second year, pages 2005 and 2006.

True veins of small granular granite * occur at *Johan-georgenstadt*, and *Eibenstock* †, where they are improperly called *sandstone veins*. At Johangeorgenstadt, they occur in a rock of mica slate, which is fine slaty, and are traversed and deranged by all the veins of silver. I have chiefly remarked this at *Glckkenklang* and *Treue-Fründschaft*. This also shows that the silver veins are of newer formation. Large veins of porphyry occur at *Marienberg*, chiefly at *Bobershau*. They are known by the name of sandstone veins. I have found, amongst others, veins of basalt in the *Plauischen-Grund*.

Veins of wacke occur in prodigious number in our *Erzgebirge*; chiefly at *Annaberg*, *Wiesenthal*, and *Joachimsthal*; they traverse all the other vein-formations, and are consequently newer. I have met with veins of greenstone at *Bauzen*, not only in the neighbourhood of the *Spree*, but also in a quarry, near the town, on the road to *Görliz*.

§ 50.

---

* It follows that this is a granite of newer formation.

† Mineralogische Geographie der Chursachsischen Lande, pages 261, and 390.

§ 50.

SEVENTH PROOF.

The mode assigned for the formation of the spaces now occupied by veins, is still farther demonstrated by the relation which veins have to one another. As

Their intersecting one another;

Their shifting one another;

Their spliting one another into branches;

Their joining and accompanying one another;

Their cutting off one another.

All these peculiarities are produced by the *effects of a new fissure upon one that is older*, which had been already filled up in part, or in whole. It is very easy to account for all this on the explanation already given of their formation, but on no other hypothesis.

When one vein intersects another (which is always the case when two veins cross), I give the following explanation of it. There existed already in the rock, a vein (that is to say, a fissure entirely filled up and closed,) at the time it was rent again; this new fissure has a different direction from the former one, is continued across it, intersects it, afterwards becomes filled with materials of a subsequent formation,

tion, and thus forms a newer vein. The last or newer vein always traverses the first or older; and the older or first formed is traversed by the last or newer, as has been already explained at § 32.

If part of a rock which has been thus intersected by the new fissure, sinks down, or suffers any derangement, and carries along with it that portion of the vein inclosed in its substance, it will not be found in the prolongation of that line which denotes the direction of that portion of the vein contained in the rock which has remained fixed. When this happens, the new vein is said to derange or alter the direction of the older.

If there were a fissure, either wholly, or in part empty, when the rock has been rent a second time in a direction which traverses the original one, then the roof and walls of the new fissure are interrupted and cut by the older, and not being able to support the pressure arising from their own weight, they have been split; so that a number of large pieces or large tables are detached, fall into the interior of the fissure, contract it, and leave new spaces between themselves and the walls of the fissure: these new spaces form so many lateral fissures. Finally, when these are filled up and closed, and consequently become veins, the new vein, where they occur, appears to be divided; in the language of miners, the vein is then said to *ramify*, or that it is composed of fragments, or that in such a place it has been broken into fragments by such a vein. The new vein may produce this ramification in the older one, if it be still open, as well as this last produces it in that which is of newer formation.

When a new fissure in a rock extends to a vein which pre-existed in it, continues in it for some way, stretches through it into the adjoining strata, and is afterwards filled with mineral matter, it also becomes a vein; and in this case the veins are said to *join*, (*sich schaaren*, or *sich anschaaren*).

When a vein continues its direction close by the side of another, whether this be through their whole extent, or only for a short way, the two veins are said to accompany each other, (*sich schleppen*).

Lastly, when a vein already formed, either by the solidity or tenacity of its substance, prevents a new fissure from extending farther and stops its course, then the former is said to cut off, arrest, or intercept the latter, (*abschneiden*).

§ 51.

§ 51.

EIGHTH PROOF.

The relation which veins bear to the rocks and beds in which they occur, or the manner in which they are found in them, proves still farther that they have been fissures. When a vein traverses the strata of a mountain, it happens almost always that the corresponding strata on its hanging side are found much lower down than on its lying side; this difference of level in the parts of the same stratum or bed, bears a proportion to the size of the vein, being always greatest in those which are of most considerable magnitude. This peculiarity is most obvious in rocks whose various beds differ much from each other in colour and external appearances. This phenomenon merits particular attention from miners, as it often occasions much embarrassment in practice.

Better marked examples of this cannot be seen than at Zinnwald. Where the tin beds of that country are traversed by veins, it often happens that the part of a bed which corresponds to the hanging side is much lower than that on the lying side of the vein, and this difference is always greater in large than small veins.

 This

This peculiarity presents itself very frequently, and in a very striking manner, at *Saalfeld.* The mountains of bituminous marl slate which are found in that place, contain a number of veins, where, at present, considerable works are carried on, and some of them to great extent. These mountains seem to rest upon the primitive mountains called *Fichtelgebirge*, which are about a league to the south, and appear to form the foot of that range. The Fichtelgebirge has here obtained the name of Thüringerwald, and is of considerable height. The veins which traverse the *Floetz rocks* in *Saalfeld* have nearly the same direction with the primitive mountains, so that they seem almost parallel to them: they have the same inclination as that of the mountain; and the beds of the rock on the hanging side are considerably lower than on the lying side, and this difference is in proportion to the magnitude of the veins. This is an incontestable proof that veins were formerly fissures produced by the weight of the mass of the rock, which, giving way on that side where it met with least resistance, has sunk down and been rent in different directions. Several fissures in the neighbourhood of these veins are still open, from which there proceeds a natural current of air, in so strong a blast as to extinguish the lights carried by those who

who pass before them; in some places they serve the purpose of galleries for running off the waters.

We observe in the coal mines near Dresden, beds, traversed by veins, that present the same phenomena as those of Zinnwald and Saalfeld.

§ 52.

NINTH PROOF.

If we examine with attention the interior structure of veins that are composed of different kinds of minerals, we perceive them to have been originally open fissures, which have been afterwards filled by degrees. Such veins are composed of beds arranged in a direction parallel to their sides; their crystallizations show these beds to have been deposited successively on each other, and that those next the walls have been first formed.

I have observed this structure in many, indeed in the greatest number of veins. In the district of Freyberg it occurs in a very striking manner in the veins *Segen-Gottes* at *Gersdorf*, in *Gregorius*, shooting from the *Alter-grüner-Zweig*, and in the agate vein at the *Corallenbruche* near *Conradsdorf*. I have in my possession a specimen from *Segen-Gottes* at *Gersdorf*, in which, reckoning from the middle, (which is composed of two beds of calc spar, in which small druses occur

here

here and there), thirteen beds of different minerals are arranged in the same order on each side of the vein; these are fluor-spar, calc-spar, heavy-spar, lead-glance, &c. In the southern vein Gregorius, the two beds which adhere to the sides of the vein are composed of crystallized quartz; next to that, on each side, is a bed of black blende mixed with iron-pyrites; this is followed by lead-glance, brown-spar, lead-glance, grey silver ore, red silver ore, silver-glance; the central part, which of course is most recently formed, is of calc-spar. It sometimes happens that one or more of the different beds are wanting.

CHAP. V.

FARTHER ELUCIDATIONS OF THE PROOFS BROUGHT FORWARD, AND OF THE THEORY RESULTING FROM THEM; AND A REFUTATION OF SOME OBJECTIONS MADE AGAINST IT.

§ 53.

IN order to form a more correct idea of the theory which has been given in the two preceding chapters, on the *formation of the cavities now occupied by veins*, we must keep it in remembrance, that several rents constituting the spaces in which veins are now found, were larger than at present, and have been since contracted; that others have become larger and larger, perhaps during the very time they were filling up; that the greater number of the old rents were already filled up, and consequently shut, at the time new ones, either having the direction of the old ones,

ones, or traversing them, were formed; and finally, that similar events may be often repeated.

§ 54.

In veins which are wide, and of considerable size, particularly when their inclination approaches to the horizontal, the hanging side, when not well supported, must yield to the pressure of the superincumbent weight: such a sinking of the strata will produce many rents and fissures, which terminating in the principal rent, will form collateral ones.

It is very probable that the veins *Freudenstein*, *Isaac*, and others, terminating in the large vein *Halsbrückner Spath*, which run in the rock forming the hanging side of the vein, and are filled with the same materials, are only collateral rents produced by the sinking down of the hanging side of the principal vein. When the hanging side of a vein, yielding thus to its own weight, falls down and approaches the walls, it must of course diminish the size of the original fissure. We often meet with such collateral fissures on the small scale; in narrow veins we can often follow them to their termination, which I have done on many occasions.

§ 55.

§ 55.

We meet with distinct examples of new veins formed in the direction of those of an older date, (either within their substance, or by the side of them), forming with them the same individual substance. We see a remarkable instance of this at Rothenberg near Schwarzenberg, in the vein *Johannis*, where a part of the substance of the vein, very distinct from the rest, is called the *gelben* and *rothen trume*, (the yellow and red branch); at Marienberg, the *Einhörner-gang* consists of two distinct masses, the one composed of ores of tin, the other of silver ore; and at Freyberg, in the Hohe-Bircker vein, with its white and red branch, (weissen und rothen trume), and the Abraham-Spath vein, in the mine Neue Morgenstern, with its *groben und spath-gänge*, the same thing occurs.

§ 56.

The difference, sometimes very considerable, in the width and size of the same vein, may proceed from the walls or roof of a fissure which runs in a curved direction, having sunk down or suffered some other derangement, by which a concavity has been placed

placed over against a concavity, and a convexity against a convexity; or large pieces of rock may have been detached from the hanging or lying sides, which have increased the size of the fissure in these places, and narrowed it in those where they have been stopt.

§ 57.

We can determine with almost mechanical precision, the position and direction of the force which has rent and produced the cavities which are now occupied by veins. For if we consider attentively the inclination and direction of the principal veins of the same formation in any one country, which are usually almost parallel, but still more those of each particular vein; we shall be able to determine with sufficient accuracy, the place from whence the force which has cut the rock has proceeded, and also the direction which it has taken. This force was nothing else but the weight of a considerable part of the rock itself, which had not found sufficient support. It is,

1. Necessary, that the force which has produced the rent, should have existed in that part of the rock which composes its hanging side.

2. This force, (that is to say, the pressure arising from

from the weight of a mass which was increasing, or which had not a sufficient support, or which had in part lost that support), has acted by forcing from the upper part of the rent a portion of the rock, and thrust it to that side which was most free, and had the least support.

3. The direction of the force which produced the interruption of continuity, passed through the centre of gravity of the mass which acted by pressure, or rather of that mass which was separated: we can easily conceive this force to act in a plane passing through the middle of the roof of the rent produced, and perpendicular to the principal line of its direction; consequently this plane must have also passed through the line of the inclination.

It would perhaps be possible to describe a line in this plane which would represent pretty nearly the direction of the force.

§ 58.

I come now to answer the objections which have, by some mineralogists, been urged against that theory of the formation of veins which considers them to have been rents. What, among other things, induces them to entertain some doubt of this mode of formation, is the intimate connection which at times takes place between different veins, and the rock which includes them. Such geognosts also reproach me with the occurrence of fissures crossing veins of a large size. But it is an easy matter to repel these specious objections, by explaining this phenomenon in a different way.

The union between a vein and the rock, on some occasions, is so intimate as to give the appearance of their having been melted together, if I may so express myself. This union, I say, proceeds from their homogeneous nature, and from the newness of their formation. In places where this peculiarity occurs, the rock has had a strong attraction for the substance of the vein introduced into the rent, and has become so intimately mixed with it, that they now appear to be one and the same substance, at least it is not easy to mark a line of separation between the rock and the vein. This is particularly the case with veins of quartz and horn-blende, when they occur in newer gneiss of a quartzy nature; but veins of pyrites in this rock do not present this appearance, which is upon the whole a rare occurrence. In general the vein and rock are very distinctly separated from each other, and there are sometimes interposed between them thin layers of an earthy matter called *besteg*. A vein is very seldom united to the rock

rock so as to adhere intimately with it through its whole course; but this only takes place in certain parts.

§ 59.

The transverse fissures which we find in veins are upon the whole but a rare occurrence, and evidently anise from the following causes. After veins have been completely formed, they have suffered violent shocks, which, pursuing the direction of the strata of the rock, cross the vein. The natural effect of these shocks acting unequally on the substance of the vein, has been to produce rents and the transverse fissures in question.

§ 60.

I cannot conceive how the great size or thickness of certain veins should be regarded as a reason against the formation of veins which has been advanced.

But if it were so, we have only to consider, that the largest veins taken alone, (that is to say, taken without others that accompany them), and without their branches, do not, in their ordinary size, exceed three fathoms in thickness. How small is this thickness when compared with the immense volume of the mass of the mountains in which they occur.

We

We rarely meet with true veins, which, taken singly, exceed a fathom in thickness. In the mountains of Freyberg, among a hundred veins, I hardly know one which is so large. In the prodigious quantity of veins which I have seen in Saxony and elsewhere, I have not hitherto found one, whose ordinary size is three fathoms; it must be understood that I do not include branches or veins.

In estimating the thickness of a vein, we must not include the adjacent rock, although it be in part decomposed or impregnated with the mineral. Nor is it proper to measure the size of a vein in those places where it is of more than ordinary thickness, or where it branches out. Still less are we, in estimating the size of a vein, to include the accompanying branches; for in these different ways we may find veins many fathoms thick.

The famous druse or cavern at Joachimsthal affords an example of the prodigious size produced by the junctions and intersections of many veins, particularly when considerable pieces of the rock are separated by these intersections. This singular cavity, which is closed on all sides, was found at the *Hohetanner Gruben-feld*, in driving the fifth gallery in the rich vein of silver *Andreas*, at the depth of 250 fathoms; this cavern (from which a large quantity of water

water flowed), was, according to report, eleven fathoms long, nine wide, and its depth, which is not yet known, exceeded twelve fathoms. I was unable to advance into this cavern further than about two fathoms; because the bottom was filled with rubbish, and the numerous pieces of rock which threaten to fall, rendered a farther passage so dangerous, that no one is permitted to attempt it. I could, however, very well see that it has the size ascribed to it. I remarked farther, that it had a long flattened form like a vein; that its inclination was almost vertical; that the rock which formed its walls and roof was quite full of rents; from these proceeded the quantity of rubbish and stones with which it was filled. In short, many veins, and among the rest, one of wacke, traverse it, joining and crossing each other at that place. From which, it appears to me that this cavern had no other origin than that which I have ascribed to it. The rock in which it occurs, as well as I can recollect, is a kind of clay slate approaching to mica slate. Ferber gives a very particular description of it in his Mineralogical Geography of Bohemia *.

Veins

* Joh. Jac. Ferber Beitrage zu der Mineralgeschichte von Bohmen, 8vo. Berlin 1774, pages 74 and 75.

Veins of an extraordinary size are said to be found in the Upper Hartz, and at Schemnitz in Lower Hungary. The *Burgstäder* vein is supposed to be the largest in the Hartz. Lasius says, that in several places it is from 20 to 30 fathoms wide *, but he had before remarked that it is rather to be considered as an assemblage of a number of veins than as a single one. He is decidedly of opinion that the large repository or block of mineral in the Rammelsberg is not a vein; it is probable that this block has been deposited and formed as a block or separate portion of the rock. With respect to the vein called the *Spitaler Hauptgang*, the largest of the veins at Schemnitz, Born says, in the 183d page of his letters, that in the Pacher-stolner

* In the second part of the work already quoted, he speaks of it at pages 305, 306, in the following passage. " The country between Mildeman and Hirscheler Teiche, " near Grube Caroline, may be considered as an assem- " blage or tissue of large veins, or it may be regarded as " one single vein which varies much in thickness: some " of these veins extend a mile in length. The vein is " sometimes extremely narrow, as is the case at the place " where the mining towns of Clausthal and Zellerfeld " meet; at other places it is from 20 to 30 fathoms wide, " as at Burgstadt, Stuffenthaler, and Zellerfeld."

stolner Feld, the place where it is largest, it may be 14 fathoms wide, and 18 if the wedges or inclosed portions of the rock be included. Thus, this vein, where it is of the greatest size, is also an assemblage of branches. But I am not convinced that the three large veins of Schemnitz are true veins. Many circumstances, and among others, the similarity of their direction and inclination, their great horizontality, and their great thickness, make me presume that they are metalliferous beds.

Besides, when I come to reflect on the immense extent of the mass of the mountains, and the extraordinary force which must sometimes be produced by their weight, I should be astonished that there have not been larger veins than those I have just mentioned, if I did not find a certain cause for it, which is the following. The fissures in which veins are formed, were, to all appearance, formerly much larger; but so long as they remained open, the rock, principally that composing their roof, yielding to the pressure of the superincumbent mass, must of necessity have narrowed the fissures, and reduced them to the small size which veins now have.

§ 61.

§ 61.

Lastly, I cannot think that any one, after reading and maturely reflecting on what has been said, will make the following objection to the theory which has been proposed to account for the formation of veins; viz. "that when two veins converging in their inclination, cut each other, and are at a certain distance traversed, at right angles, by two other veins that are distant from one another, then a portion of the rock of a prismatic form is isolated and separated from the rest of the mass of the mountain. Now, it may be said, if veins have originally been empty fissures, this portion of the rock, deprived of every support, must have been suspended free and unattached, till such time as the matter which composed the vein, and which surrounds this prism, had filled, in part, at least, the fissures, which was absolutely impossible." I shall once for all answer such an objection, by saying, that when a theory is to be combated by facts, it is necessary, in the first instance, to prove these facts, and then to mention where they are to be met with. Otherwise facts very contrary to the nature of things may be supposed, or mentioned, which have no existence, and cannot be made

use

use of to combat an explanation. Such is precisely the case with this objection, which supposes fissures that have different directions and dips, and which cross each other, to have been formed at the same time. Wherever I have seen veins of considerable size crossing each other, I have always found that they had been formed at different periods. For when two veins of different directions and inclinations meet, one of them always intersects the other; this may in its turn be traversed by a third, this third by a fourth of latter formation, and so on: in this way we see that the first rent had been filled before it was traversed by the second, and that this in like manner was filled before the third was formed. I have inquired, respecting the manner in which veins occur in other mountains, at different observers acquainted with them; and have always been informed, that when two veins cross, one of them always traverses the other, and is consequently of later formation.

It is besides natural to suppose, that when a mountain is rent at several different times, it has been done at each of these, by one and the same force acting in the same direction; so that all the veins which have been formed at the same time ought to be parallel, or nearly so. If this be really the case, then, when veins cross several times, or when they converge, they must have been produced at different epochs; in which case there is no difficulty in regarding these veins as having been originally open fissures.

We might, if the question were only about small veins, admit the objection; and suppose that veins, which in their intersections isolate a portion of the rock, have been produced at the same time. But this cannot contradict the mode of formation which we have assigned to those spaces now filled by veins, viz. that they have been open fissures in the rock. In fact, it is possible to conceive that a portion of the rock, separated from the mass of the mountain by fissures which surround it, might support itself until they were filled up with mineral matter, and till this matter was consolidated. For this purpose we have only to suppose, that when fissures are formed, there are always detached from the rocks some fragments, which, falling into these fissures, will prevent the contact of their walls, and which will in consequence be kept separated from each other. Besides, the walls of a rent are not in general plain and smooth, but rugged and covered with inequalities: thus, as soon as a fissure is made, if one of the parts of the rock that is rent has suffered any impulse, it may very easily happen that some of these inequali-

ties

ties may be opposed to others, which will prevent the divided parts of the rock from re-uniting, and thus form an open space between them. Thus it is very possible to suppose that veins which surround a portion of rock have been fissures, all of which were formed at the same time, and which, in part at least, were open.

CHAP.

CHAP. VI.

PROOFS, THAT THE SPACES WHICH VEINS NOW OCCUPY, HAVE BEEN FILLED FROM ABOVE.

§ 62.

I NOW come to the consideration of the proofs of the second grand proposition of my theory; namely, the following.

The mass of veins has been formed by a series of precipitations, which have filled in whole, or in part, the spaces now occupied by veins: these have entered by the superior parts of the rents which were open; and have been furnished by a solution in water, generally chemical, which covered the country in which these rents then existed.

I shall adduce only three proofs in support of this proposition; and they will, I think, be sufficient to establish it entirely.

I lay down, in the outset, the following proposition in geognosy, to show the incontestability of these proofs, more especially the first of them; it is evident, and universally received.

All the flöetz mountains, as well as all the others which resemble them in their stratified structure, and in the nature of their integrant parts; all such mountains, I say, are formed by an assemblage of sediments and precipitates proceeding from the waters which covered the globe. Each of these depositions has formed a particular bed; and all these beds, as we now see them accumulated on one another, are, beginning from below, a succession of precipitates which have been formed after each other, or rather deposited upon one another.

In order to understand perfectly, and be able to judge of the theory which I am going to give, in its application to the manner in which veins have been filled, or to the formation of the substance composing them, it will be necessary to have a just conception of the difference between a chemical precipitate and a mechanical deposition: to have a perfect idea of simple elementary bodies which are not susceptible of any transmutation; to be acquainted with the theory of solution and precipitation founded upon chemical affinities; and above all, to know that the same individual

dual solution may furnish, not only at the same time, but also in succession, precipitates of a different nature.

§ 63.

FIRST PROOF.

When mountains and whole tracts of country, in which open fissures existed, were covered by the waters holding different substances in solution, and when these were precipitated, it must of necessity have happened, that these precipitates would enter into, and be deposited in the fissures. All the beds and strata of flöetz mountains, as well as of primitive mountains, are precipitates; and these precipitates (almost all chemical, and formed in the humid way), were furnished by solutions that covered the tracts of country where beds and strata are found, and even extended to a much greater distance. As I have before said, fissures were, from time to time, formed in the mountains, but chiefly about the period of their first existence; so that at different times there existed fissures which were entirely empty and open, or partly filled up. Whilst these fissures were covered with the solutions, which, by successive precipitations, have formed the beds and strata of the primitive and flöetz mountains, the different materials (the nature of which varied

ried at different times), contained in the solution, were precipitated in the fissures which they thus gradually filled up: the nature of these substances depended entirely on the kind of the precipitate which formed them, and consequently also on that of the solution at the time they were formed from it. This is the reason why we find in veins nearly the same fossils which form the beds and strata of mountains.

Three important causes, which it is necessary not to lose sight of in examining this theory, may have occasioned some difference between the matter of veins and that of beds. 1. The precipitations and depositions which formed veins, have been made more tranquilly than those which produced beds. 2. Mechanical solutions and depositions have disturbed the formation of veins much less than that of beds. The great number of crystals, and the nature of fossils found in veins, show the formation of them to have been more tranquil, more slow, and made more freely. 3. The spaces in which veins are formed, have preserved for a longer time the faculty of receiving and retaining different solutions; and when one solution was exhausted, they may have received another: in this way veins often contain fossils of different formations, whilst the beds of a mountain

mountain contain only those of one formation; these last are therefore more uniform in their mass than veins.

§ 64.

If we compare the substances, which, under the form of precipitates, constitute the mass of mountains with the mass of veins, we will in general, if not always, find a very striking resemblance between them. We find, as has already been remarked, at Johangeorgenstadt, veins of granite of newer formation; at *Marienberg*, veins filled with newer porphyry; at *Wehrau*, in Upper Lusatia, there are veins of coal; and at *Aehlen*, in the canton of Berne, we find veins of rock salt. That great revolution of nature which produced the mountains of the trap formation, has in all likelihood occasioned a vast number of rents; thus we find, in almost all mountains, veins filled with the materials which constitute the rocks of the trap formation, as basalt, wacke, greenstone, and amygdaloid. Lastly, It is very well known that quartz, calc-spar, and clay alone, constitute veins.

§ 65.

I have just shown that fossils which are commonly found constituting the beds and substance of whole mountains, also constitute the mass of veins: I shall now show that the greater part of the other fossils which form veins, are also found in beds. Lead glance occurs in beds; as we see in our *Erzgebirge*, at *Geier* and *Schwartzenberg;* in Sweden, and several other places; in the mountains of Cracow and the Ardennes.

Tin ore is found in beds in our Erzgebirge, at Zinnwald and other places, and at *Gieren* in Lower Silesia.

All the copper ores are found in beds, in primitive and other rocks. They occur in beds in our Erzgebirge at Gishübel; in Bohemia at Kupferberg; in Silesia, the Bannat, Upper Hungary, Sweden and Norway. They occur in the newer rocks in the county of Mansfeld, Thuringia; in the mountains of Cracow, at Medziana-Gora; in the Uralian mountains in Russia, and in several other places.

Brown ironstone and sparry ironstone, which occur, particularly the last, so often in veins, are to be found at Kamsdorf, Schmalkalden, Eisenerz and elsewhere, sometimes in beds, and sometimes in mountain masses.

Lest I should go too far, I will not dwell longer on this subject, for I could produce similar examples from all the metals. Arsenic pyrites, blende, native gold, cinnabar, and many others, occur frequently in beds. I must, however, further remark, that the heavy spar, and fluor spar, so often met with in veins, are also found in beds of considerable size; the the former in Savoy, the latter in the Thuringer Wald.

§ 66.

SECOND PROOF.

When veins are filled with rolled pieces, as (§ 44.) is the case at Joachimsthal, and other places, and when veins contain petrifactions; it is not possible to suppose that they could have gained admission into the vein in any other way than from above.

§ 67.

THIRD PROOF.

The greater number of veins composed of different fossils, are, as has already been observed, (§ 54.), formed

formed by a collection of beds parallel to their walls. These beds are arranged in the same determinate order on each side; the corresponding beds of both sides being of the same nature and thickness. The beds on the outer part, i. e. those that touch the walls, are thinnest towards the top, become thick in proportion as they descend, and still lower they sometimes terminate by meeting and joining together. The crystals, and their impressions, show that the beds next the walls were first formed; the others followed in succession, finishing with those in the centre, in which we often find druses.

I have already remarked that I have in my possession three specimens of veins of the Freyberg district, in which the formation and structure spoken of are distinctly marked. Is it possible to account for this order and regularity in any other way than by supposing, that the spaces in which veins are formed have been filled by a chemical solution, that the nature of these solutions varied at different times, and that from them the different beds composing the vein were successively formed by precipitates arranged upon one another? The first precipitation having covered with solid matter the sides and bottom of the cavity in which the vein is formed, and having,

having, as it were, closed and shut up the orifices of the different pretended openings which terminated in the walls of the rent, how can we imagine the successive solutions to have entered, but from above?

CHAP.

CHAP. VII.

REPLY TO THE OBJECTIONS WHICH HAVE BEEN, AND MAY BE URGED, AGAINST THE EXPLANATION THAT HAS BEEN GIVEN OF THE MANNER IN WHICH VEINS WERE FILLED UP: AND CONTINUATION OF THE ACCOUNT OF MY THEORY FOR THE FORMATION OF VEINS IN GENERAL; AND, IN PARTICULAR, OF THE MANNER IN WHICH THEY HAVE BEEN FILLED WITH MINERAL MATTERS.

§ 68.

It will perhaps be asked from what source the metallic particles, and all the mineral matters which were contained in the vast solutions or waters which overspread whole countries, were derived. Some will imagine, that, in proposing this question, a very great objection is made to what I have just said on the manner in which veins have been filled up. To this

this I reply, that although we do not know from whence these particles have come, this circumstance does not prevent us from perceiving the existence of a phenomenon, which, with all its consequences, is before our eyes. Of the state of ignorance in which we remain, with regard to the origin of the metallic and mineral matters, we never can avail ourselves, as a means of combating the fact itself. In the mean time, we must content ourselves with knowing, that, at certain periods, the materials which now constitute the substance of veins, were in reality contained in the sea which covered our globe universally: and we must wait with patience till new observations teach us (if it be possible), from whence the component particles were derived, and by what means they were introduced into the general solvent. In all researches into the effects of nature, and their causes, as well proximate as remote, we at last arrive at the investigation of ultimate causes, beyond which we cannot proceed. In some cases it is even difficult to discover the remote cause of certain effects and phenomena.

In recapitulating the state of our present knowledge, it is obvious that we know with certainty, that the flöetz and primitive mountains have been produced by a series of precipitations and depositions formed

formed in succession; that these took place from water which covered the globe, existing always more or less generally, and containing the different substances which have been produced from them, (§ 30. and 62.). We are also certain that the fossils which constitute the beds and strata of mountains were dissolved in this universal water, and were precipitated from it: consequently the metals and minerals found in primitive rocks, and in the beds of flöetz mountains, were also contained in this universal solvent, and that they also were formed from it by precipitation, (§ 40. and 62.). We are still farther certain, that, at different periods, different fossils have been formed from it, at one time earthy, at another metallic minerals, at a third time other fossils, (§ 62. and 63.). We know too, from the position of these fossils, one above another, to determine with the utmost precision, which are the oldest, and which the newest precipitates (§ 62.). We are also convinced, that the solid mass of our globe has been produced by a series of precipitations formed in succession, (in the humid way); that the pressure of the materials, thus accumulated, was not the same throughout the whole; and that this difference of pressure, and several other concurring causes, have produced rents in the substance of the earth, chiefly in the most elevated

vated parts of its surface, (§ 39. and 40.). We are also persuaded, that the precipitates, taking place from the universal water, must have entered into the open fissures, which the water covered (§ 30. and 63.). We know, moreover, for certain, that veins bear all the marks of fissures formed at different times; and, by the causes which have been assigned for their formation, (§ 41. 42. 50.), that the mass of veins is absolutely of the same nature as the beds and strata of mountains (§ 64. 65.), and that the nature of the masses differs only according to the locality of the cavity where they occur, (§ 63.). In fact, the solution contained in its great reservoir (that excavation which held the universal water) was necessarily subjected to a variety of motion, whilst that part of it which was confined to the fissures was undisturbed, and deposited, in a state of tranquillity, its precipitate.—Further, from time to time there were diffused, in the general solution, to a great distance, mechanical diffusions, which formed precipitations of great extent, and these have been mixed with the chemical precipitate; but as these mechanical solutions did not penetrate into the fissures, except in small quantity, they could not disturb or alter the precipitation there going on, and from which the mass of the

vein

vein was formed in a state of tranquillity. The precipitates which formed the beds of mountains, have, of necessity, deposited on the bottom of the general reservoir, solid and compact materials; whilst the matter which composed the greater part of the mass of veins, being deposited by degrees on their walls, has there formed druses: Afterwards, minerals of different natures have been successively deposited upon one another. These precipitations and consequent crystallizations, shortened, narrowed, and even sometimes completely filled up the drusy cavities which formerly existed (§ 43. 52. 67.).

In short, the beds of mountains, formed in the manner just noticed, containing fossils of but one formation, must of course be very simple. Veins, on the contrary, contained, principally in their upper parts, druses, which thus afforded room for the reception of subsequent formations. The solutions containing the formations, being received into these cavities, deposited their contents in them: from this arises the great variety which the substances of veins present. We know for certain, that veins have been formed at very different times (§ 31. 50. 53.); and that we can not only determine the different formations, but even ascertain the relative age of each, (§ 51. and 53.). Lastly, we know for certain, that new veins have been formed in countries where old ones occur (§ 50. and 53.). These new ones are the cause of all the peculiarities presented by veins, in their intersections, meetings, junctions and derangements (§ 50. 51.). We know that in some veins fissures have been formed, sometimes on their sides, and sometimes in their middle; which new fissures have been filled by subsequent formations (§ 36. 53. 63.): in this way are produced the varieties, which the substance of veins, compared with beds, exhibits.

Such are the principal propositions of the new theory of veins, propositions which have been demonstrated in this treatise; they are connected with each other, they proceed from one another, and they are founded upon and proved by the observations which have been brought forward. It appears to me that this theory is a great step gained in the knowledge of the nature of veins, and consequently in the natural history of our globe, or in geognosy. Every application which may be made of the theory of veins to the working of mines, arises from the propositions laid down in the short recapitulation of it which has just been given. In the ninth chapter of this treatise, I shall give more in detail the application of this theory to the practice of mining. The

practical

practical miner who is most interested in an exact knowledge of the nature of veins, may be satisfied with these principal propositions, their consequences and application; they may be considered by him as a sufficient introduction to a more exact and careful study of veins, and of their properties, both local and particular.

Nevertheless, the geognost, though convinced that the integrant particles of strata, beds, and veins, were contained in the universal waters, will ask not only from whence they have been derived, but also at what period they were introduced into them? The geognost, who is possessed of the necessary knowledge of chemistry, and consequently of the impossibility of one elementary substance being transmuted into another, will see that there are only two ways in which the following question can be answered. At what time the metallic, earthy, and other substances, which were, and still are in part, contained in the general solution, and which have formed depositions not only in the bottom of the reservoir, but also in the fissures of rocks which they have filled; at what time, I say, have these substances entered into the general solution? It may be answered, either that these substances have altogether, and from the beginning, been contained in the

the universal solvent, or that they may have been introduced from time to time, one after another, and that they may have been in it, in greater or less number, at the same time; and if we admit the first answer, it is not possible to understand, why, since the solution contained from the beginning all the materials which have since been precipitated from it, why, I say, at different periods, successive depositions should have been formed of so different a nature, without our being able to observe between two consecutive precipitates, something that might show one of these to have remained longer in solution than the other, or that the deposition of the first had necessarily occasioned that of the second. Thus, it is not possible to conceive, why, in a mountain of gneiss, the strata of this rock should alternate many times with beds, in some instances, of limestone, sometimes of hornblende, lead-glance, and other metallic minerals; sometimes of magnetic ironstone, quartz, felspar, &c.; all of which are essentially different from gneiss: sometimes also of limestone, clay, marl, lead-glance with calamine, chalk and flint; and this perhaps for more than a hundred times: how does it happen that these mineral substances are precipitated alternately and in succession, yet in such a manner, that most commonly we do not find in any one of the beds the

least

least vestige of that which preceded it? It is therefore most probable, that at different periods the universal solvent contained mixtures as various as the different precipitates; and that the universal waters held in solution at one time one substance, and at another, another; in a word, that at different periods, different substances have entered into, and been retained in the solution.

From what has been said in this section, it must be obvious, that the natural history of veins cannot be thoroughly understood without a knowledge of the primitive and flöetz rocks, as well as of their mode of formation. And at the same time that an acquaintance with the natural history of rocks throws a great light on that of veins, just as the natural history of veins elucidates in its turn that of rocks. To study this subject completely, we must have sufficient knowledge, not only of the different kinds of mountains, of their rocks, of the peculiarities of their structure, both generally and particularly, principally with respect to the different kinds of formation; of the relative age of these different formations, as well as of that of the rocks of intermediate formation, but also of the peculiarities of the structure of rocks, that is to say, of their stratification and super-position, from which alone we can determine their relative

tive age. In studying more particularly the different rock formations, we must begin with the newest, which are the alluvial; and from these, ascend successively to the most ancient. From the alluvial we pass to the newest flöetz mountains, and so on through the transition to the oldest primitive mountains. The object and limits of this treatise will not permit me to enter more at large on this subject. In detailing my theory of veins, I must suppose my readers to have a knowledge of rocks, which is indispensably necessary to be able to form an opinion of the theory.

§ 69.

The following objection cannot now be urged against my theory; that when veins contain at the same time different kinds of vein-stones and ores, it is difficult to conceive from whence these fossils have come, which are often found in the same vein, and how the solutions from which they were produced could be preserved so distinct. I have sufficiently shown, in the preceding passages, that this variety in the fossils of which a vein is composed, arises in part from the substances which form it, having been introduced into the fissures at different periods; that the solution which yielded the precipi-

tates

tates contained different substances at different times (§ 63. 68.); and that a part of the different fossils which are seen in the same vein, were in fact precipitated from one and the same solution, some sooner, others later, and a few even at the same time (§ 62.). Further, chemistry shows, that the same solution may contain very different substances; and that in a compound solution, there may be formed not only at the same time, but in succession, precipitates of a very different nature. From what has already been delivered on the structure of veins, and the locality of different fossils that are found in the same vein, and still farther what will be said in giving a description of the depot of metalliferous veins in the district of Freyberg, (chiefly the first and second), it will be sufficiently plain, that of the different fossils which compose a vein, one part has been formed and deposited at the same time, another has been produced by a succession of precipitates made one after another, and that the rest have been formed at times very distant from each other.

§ 70.

Perhaps it may be objected to the new theory, that, as has already been remarked, fragments of rocks are found in the body of a vein (§. 45.). It

may

may perhaps be difficult to understand how the fragments of a rock can be supported in an empty space, or they may be considered as proofs of the theory of transmutation, that is, of the transmutation of the rock into the matter of the vein. To the first of these difficulties I reply, that there are many ways in which we may imagine the fragments of the rock to have been supported in the position and place where they are now found, and that in all probability it is sometimes done in one way, and sometimes in another. For example, it may sometimes have happened that some of these fragments, after being detached from the rock, fell into the fissure, and having reached a part too narrow to allow them to pass, have been detained there. At another time, it may have happened, that after a fissure was completely filled up, another rent has been formed either in the sides or middle of the first; the shock which produced this effect may have detached pieces of the rock which fell into the fissures, and were stopt in narrow spaces or pre-existing druses. It might farther happen that, in a vein, a portion of the rock adjacent to the vein might have contracted a very strong adhesion with its mass; and if this vein were again rent on its side, it would carry along with it this adhering piece. Again, when a rent is made

produced, the walls might have been split by the shock, from which fragments would be detached, either in isolated pieces, or in numerous small fragments. When the fissures came afterwards to be filled up, all the fragments which had fallen in by the different causes already mentioned, have been surrounded with, and enveloped in, the matter composing the vein. There is also a fifth way of explaining the manner in which the fragments of rock have been supported in the vein; they may have fallen into the solution at the time it filled the fissure, and was depositing the substance of the vein; they would be supported in this matter, which was still soft, nearly in the same way as a stone thrown into a vessel where the water is freezing, is supported in the middle of the liquid which is passing into ice.

We cannot avail ourselves of this circumstance of the fragments of rocks being found in veins, as an argument in support of the theory of transmutation; for these fragments have very sharp edges and angles, which evidently show them to have been detached from the neighbouring rock. Farther, when they are small, they are confusedly mixed together, and assume all kinds of directions.

§ 71.

In most veins, the rock which forms the walls is more or less altered and decomposed. This happens chiefly when the rocks in which veins occur are of granite, sienite, gneiss, mica slate, clay slate and porphyry. In such instances, it is only one of the component parts of the rock that is decomposed; never the quartz, commonly the felspar, more frequently the hornblende, and often the mica. This change sometimes extends to a considerable depth in the substance of the rock, even to a fathom; it does not, however, prevail throughout the whole extent of the vein, but extends farther in some places than in others, and is most general in those places where the mineral contains sulphur. It extends sometimes to a considerable distance from the point of contact of the mineral with the rock; so that in following a sterile vein, when we come to a place where the rock is decomposed, we may conclude, that we will soon find the metallic mineral.

Many geognosts think, that the remarkable change of the rock adjacent to a vein, cannot be reconciled to that theory which regards veins as open fissures that have afterwards been filled up, but that it rather

ther proves the mass of the vein to have been produced by a transmutation of the substance of the rock. The remarkable difference, however, which is found between this altered rock and the mass of the vein, contradicts most completely all idea of transmutation; for the altered portion contains almost always the same parts, and in the same proportions: the action of some acid may have weakened the chemical union, and consequently diminished the cohesion of the parts; perhaps, too, an acid may have dissolved one of the constituent parts, which in this way has disappeared; but the whole may still preserve the same texture, and the difference between the adjacent rock and the substance of the vein is so considerable, that it does not admit of the most distant idea of transmutation, the smallest trace of which is nowhere to be discovered (to say nothing of the chemical impossibility of such a change). This change appears to have been effected by the action of acids which existed in the solution that formed the vein, whilst it filled the rents; these insinuated themselves into the neighbouring rocks, which they have changed in a greater or less degree; and so this phenomenon can be easily reconciled to our theory.

I have remarked in rocks two distinct kinds of decomposition,

composition, proceeding, no doubt, from the action of two different acids. First, I have found, in mountains of granite and gneiss, the felspar only decomposed and converted into a white porcelain earth, whilst the quartz and the mica have suffered no alteration. I have observed similar decompositions, not only in the walls of several veins, principally of those whose minerals contain carbonic acid, but also on the surface of mountains. This change I ascribe to the action of the carbonic acid.

Gneiss, changed in this way, is found in the walls of *Halsbrückner-spath*, near Freyberg, chiefly in the old mines *Johannes* and *Jacob*; and a granite that has undergone a similar change, is found in the walls of a vein of brown ironstone at *Spitzleute*, between *Blauenthal* and *Schneeberg*. In the suburbs of Freyberg, on the road to the *Munzbach-Hütte*, we find gneiss much decomposed on the surface of the mountain, and which penetrates a considerable way into its substance. And a similarly decomposed granite occurs on the surface of a mountain at *Burkartzgruner Seiffen*, not far from *Schneeberg*.

The second kind of decomposition of rocks occurs only in considerable veins: it principally attacks the felspar and the mica, and sometimes hornblende, when it meets with it; these substances are convert-

ed

ed into a kind of green lithomarge and steatite, which pass into another *. This last is that variety of decomposition, which at the beginning of this section I noticed. I consider it as the effects of the sulphuric acid. I have principally met with it in places which abound with iron pyrites.

In the district of Freyberg, almost all the veins of the first, second, and third formation, contain this greenish gneiss in a state of greater or less decomposition. In the *Grund* between *Freyberg* and *Dresden*, we see rocks of porphyry that have undergone this change in the neighbourhood of veins of lead-glance. The same thing occurs in a rock of sienite at *Meissen*. We also find clay slate in a similar state of decomposition near a vein at *Munzig*, between *Freyberg* and *Meissen*.

Lastly, It appears to me that the arsenical acid produces a change similar to that induced by the sulphuric. The first of these changes sometimes produces

---

* This green decomposed rock, is what was formerly called *gneiss* in the Freyberg district, whatever its original nature may have been: but that name is now applied to a particular species of rock.

produces the second, or rather they take place at the same time.

Sienite, which has undergone a decomposition intermediate between these two, but approaching more to the nature of the first, occurs at *Altenberg* in the *Neufänger-Gebirge*, near the *Stockwerke*, in the walls of a tin vein. Pieces of this decomposed sienite are found on the old rubbish, or mining heaps.

§ 72.

The antagonists of this theory think they urge a strong objection against it, when they alledge, that the part of the rock adjacent to veins is often impregnated with minerals; and this circumstance appears to them the strongest proof that can be adduced in support of the opinion which supposes the rock to have been transformed into the matter of the vein. Yet this phenomenon can be easily explained by my theory. In fact, this peculiarity occurs almost always in disintegrated rocks, and chiefly when they are porous, full of chinks and slaty, and proceeds from the attraction of the rock for the metallic particles contained in the solution that filled the rents in which the veins were formed; or rather, it it proceeds from the attraction which the rock had for the solution containing the constituent parts of minerals, which had insinuated itself into the substance of the rock.

Most commonly the mineral found in the rock occurs under a *superficial form*, (i. e. in thin leaves), which is generally the same shape with that which occurs in rents in the mass of a vein, particularly in those which are between the vein and the rock. In this case, these minerals are newer than the rest of the vein. Sometimes the mineral is disseminated through the rock in small grains, and then it is of the same formation as the vein. This property which some ores possess of penetrating into the empty spaces and fissures of the rock, which terminate in a vein, is peculiar to certain varieties, chiefly to native silver, silver-glance, red silver ore, native copper, tin ore, iron pyrites, and red iron ochre; it seldom happens with copper pyrites and lead-glance. The metallic particles which are found in rocks, seem to have been carried there by the effects of the attraction of the rock for the constituent parts of the mineral solution, which, entering the fissures of the rock, has there deposited its precipitates: in explaining this fact, we may also admit ordinary attraction, acting in the same manner as when it produces the ascent of water in capillary

tubes.

tubes. I must, however, remark, that this phenomenon is not very common, and that in those veins where it does happen, it only occurs in certain places, very rarely extending into the rock above half a fathom, most commonly only a few inches.

Red silver ore, silver-glance, and native silver, in a superficial form, occur in a decomposed gneiss which forms the walls of the veins in the mines *Alt-grün Zweig*, and *Himmelsfürst* at Freyberg; in both of these places a portion of the rock is worked and used as the mass of the vein. The mine of *Romische-Adler*, at Johangeorgenstadt, formerly contained native silver in thin leaves in great quantity, which was found in a rock of clay slate approaching to mica slate; it is still found at *Kongsberg* in Norway, in rocks of gneiss, mica slate, hornblende, &c. At *Marienberg*, in the mine *Drei-weiber*, there is a vein which yields copper-glance; the adjoining gneiss, which is decomposed, is impregnated with ferruginous particles, and contains native copper in a superficial form. Some time ago, there was found in the mine *Morgenstern*, near Freyberg, in a particular part of the roof of the vein, a decomposed gneiss enclosing small leaves of lead-glance. Gneiss, clay slate, and mica slate, are often impregnated with iron pyrites when it occurs in veins. When the

the rock which is in the vicinity of veins of red iron-stone, or of other veins which contain much red ochre, is decomposed, it is almost always found to be penetrated with the ochre. Lastly, The part of the rock adjacent to the tin veins is filled with grains of that metal; examples of which are to be seen at *Altenberg*, *Ehrenfriedersdorf*, and *Geier*. I shall afterwards (§ 74.) speak of this rock impregnated with tin.

§ 73.

A fact which has been noticed by several writers*, has been opposed to my new theory of veins, not only in respect to what has been said on their mode of formation, but also in respect of the manner in which they have been filled. It is in treating of a remarkable peculiarity which the veins at the Peak in Derbyshire present, with respect to the beds of amygdaloid found in the mountains of that country. Indeed

 this

---

* In the above mentioned work, he says, "I suppose "that the veins may have been deranged and thrown to a "side by the beds of amygdaloid, so that they may have "ramified, and that these ramifications have again united "below the bed." Vide Whitehurst's Theory of the Earth.

this fact may be urged as an objection against every theory. From what has been said on these veins, they seem to occur in floetz limestone; they begin at the surface, and traverse the rock without interruption, till they reach a large bed of amygdaloid, where their progress is stopt, and they do not cross that bed: but, underneath this bed, the limestone rock is again found, and the veins are continued through it in the same manner as above, of the same size, containing the same minerals, and having the same position; they are again stopt by a second bed of amygdaloid which comes in their way, after which they continue their course in the limestone, observing always the same thing with respect to the beds of amygdaloid they alternately meet. Before attempting any explanation of this phenomenon, I shall observe, that we must first be convinced that the facts are exactly such as they have been represented; and farther, that on a more particular examination some peculiarities may be observed, which may assist in giving an explanation of the phenomenon. It has often happened to me, that, in examining certain facts, I have found them very different from the representation which has been given of them; and, that upon examination, the difficulties have altogether disappeared. Several persons (and among the number,

number, my pupil, Mr Barker, a native of Bakewell in Derbyshire, and an eye-witness), have informed me, that in several places, the veins do actually traverse the beds of amygdaloid. The same thing has been observed by Mr Pilkington in the first part of his Natural History of Derbyshire (1790.). But supposing the fact to be as it is represented, that veins do not traverse the beds of amygdaloid: notwithstanding the great thickness and hardness which these beds have in some places, and which in others is very inconsiderable, it is not impossible but that when the rent, in extending through the rock, arrived at these thick and very compact parts, it may not have been continued; but at a certain distance, and where the bed was thinnest or less firm, it may have been traversed by the rent. Ferber, who had been informed of this phenomenon by Whitehurst, thought, that in some places the beds of toadstone were split, and that by means of these fissures, the upper part of the vein communicated with the lower. Future observations, made with accuracy, and in sufficient number, will throw farther light on this subject. It is surprising, that amid the immense number of observations made on veins, nothing similar to this has hitherto been observed in any other place.

§ 74.

Some geognosts still entertain doubts on the new theory of veins, from observing, that so little has been discovered, and is known, of the precipitates formed from the solutions which have furnished the mass of veins. We ought, say they also, frequently to meet with precipitates in the form of beds, whose substance should bear some analogy to the matter of veins. I reply to this objection, by requesting such persons to observe how scanty a stock of facts we are possessed of on this subject; how difficult it is to make observations on it in an accurate manner, in order to give positive results. It will not be, till well informed and skilful geognosts shall have made correct observations, continued for many years, in different countries, and in various parts of the world, on metals, ores, and other fossils, as well as on the manner in which they are disposed with respect to one another; and it will not be till these observations are compared together; that it will probably be found that the greatest part of the principal formations of veins (viz. the ores and vein-stones of which they consist), occur also in beds.

This

This has already been confirmed by observation, with respect at least to some minerals.

In the valley of *Lahn*, we find mineral beds of the same nature with the vein-formations which occur in the Hartz, lying between beds of grey wacke and clay slate: these consist of lead-glance, grey copper ore, brown blende, sparry ironstone and quartz. In the *Bannat of Temeswar*, the same formation is said to occur in beds, which is found in veins in *Voigtland*, the country of *Bareuth*, the *Hartz* near *Lauterberg*, and in *Westerwald*. This formation consists of copper pyrites, red ochre of copper, malachite, compact brown ironstone and quartz. The formation of liver pyrites, sparry ironstone and heavy spar, of which we see so many veins, constitutes the substance of rocks and beds at *Kamsdorf*, *Schmalkalden*, *Eizenerz* in *Stiria*, and at *Hüttenberg* in *Carinthia*. We have already remarked how the older metalliferous solutions which covered the mountains deposited minerals in the form of beds (§ 64. 65.). What a prodigious number of beds of iron and copper pyrites is there not in Sweden and Norway, which were formerly considered as veins! There are some countries in which whole ranges of mountains are entirely composed of ores; as at *Rammelsberg* near *Goslar*, and *Schlangenberg* in Siberia. In several

places

places in Saxony, metalliferous beds also occur. All the frontier of Bohemia, near *Gottesgabe*, in going from *Johangeorgenstadt* to *Breitenbrun*, *Schwartzenberg*, *Raschau*, *Elterlein*, and *Geier*, all the way to *Ehrenfriedersdorf* and *Thum*, contains an immense quantity of beds of tinstone, lead-glance, iron pyrites, and iron-ore. The country around *Gieshübel* contains also many beds of copper ore, lead-glance, and ironstone. There are many mineral repositories in our mountains, but it has not yet been ascertained whether they are beds or veins. The greater part of the mineral repositories in the county of *Gömerer*, and *Zips* in Upper Hungary, which contain copper pyrites, grey copper ore, sparry ironstone and cobalt, are (according to the account of well-informed observers, and the specimens I have seen), beds. At present I pass over many other countries. But we have only to consider with more attention several of the mineral masses that occur in beds, to compare them with the vein formations already known, and we shall in general be able to discover a strong analogy and resemblance between veins and beds.

In this investigation, it must, however, be held in remembrance, that many mineral beds which contained the same substances, as certain vein formations, may have been destroyed by nature, just as has

has happened to some of the other parts of mountains. Farther, that the formation of several beds of ores and other fossils has not been generally extended over the whole of the surface which was covered with the solution from which these precipitates have been formed. Again, that the top or surface of many mineral beds has been covered with a new formation of rocks, or else by newer strata of the primitive mountains, which renders it difficult to discover these mineral beds on the exterior of the mountains. And, lastly, that after the reasons which have been assigned (§ 63.) for the difference to be observed between the formation or composition of veins and of beds, (even when these have resulted from one and the same solution), we ought not to expect to find a perfect resemblance between them. From this we see, that without a perfect knowledge of orictognosy, it is not possible to say any thing positive, certain, or satisfactory on this subject.

§ 75.

After having replied in the preceding 68th and 74th sections, to the objections urged against my theory of the formation of veins, and after having anticipated and refuted what might afterwards be brought

against

against it, there remain still a few things to be discussed, to complete the exposition of my theory.

It is of importance for this theory, to consider not only the relative antiquity of veins and vein-formations in general (treated of § 31. 32.); but we must also attend to the age of veins relatively to the rocks in which they occur. We can in some measure determine this, by comparing together the ages of the different veins which occur in a mountain. In this way we can determine which are those whose age comes nearest to that of the mountain, or which of them differ most from it, and consequently are relatively newer than the mountain. Among the oldest veins which occur in different countries, some are comparatively much newer than the rocks which they traverse, whilst in other countries they may be nearly of the same age as the rocks in which they occur. This actually happens, and we find, (a circumstance worthy of remark), in certain places, veins which have been produced soon after the formation of the rocks which surround them, and in some measure during the formation of the mountain, that is to say, before the rock has been completely dried and consolidated. Veins, whose age approaches near to that of the rock in which they occur, are characterized by a certain analogy which they bear to the mass of the

rock,

rock, by the close connection and strong adhesion which subsists between them, and by the thinness of the vein: farther, the rock which surrounds these veins is penetrated by, and impregnated with, not only the ores, but even with the substance of the veinstones. *Geier*, *Ehrenfriedersdorf*, and *Altenberg*, present striking and distinct examples of veins, the date of whose formation is nearly the same as that of the rock. The tin veins which occur in the places just mentioned, are of this kind. We find, besides, veins of greater antiquity consisting of earthy substances, as felspar, quartz, mica, and sometimes schorl. At *Elbogen*, on the right bank of the *Eger*, close by the high road, there is a small vein of white felspar; it appears to have been formed nearly at the same time with the mountain, and occurs in a rock of granite.

When veins whose age approaches near to that of the rock in which they occur, contain ores, (take as an example the tin veins which have just been mentioned), the ore is commonly found only on the sides (Saalbander); and sometimes, almost wholly, in the adjacent rock (§ 73.); and the middle of the vein, which owes its origin to a later fissure, generally contains either stony matter, or sterile earth.

The Stockwerke (§ 4.) is nothing but a great number of small short veins collected in a greater or less space of the rock; these veins run in all directions, and their age is nearly the same as that of the rock in which they occur. As yet, I only know of one kind of ore that occurs in the Stockwerke, which is the ore of tin: repositories of this kind are found at *Altenberg*, at *Seiffen* near *Grunthenthal*, *Geier*, and at *Schlakkewald*, not far from *Carlsbad*.

§ 76.

An object of still greater importance to an observing geognost than the relation which veins bear to the mass of the rock in which they occur, is the relative age of the different fossils which constitute the substance of the vein, but more particularly the relative age of the different metals in regard to one another. The rocks in which these fossils are found, whether they occur in beds or veins, afford the greatest assistance in this research; for it is certain that those ores and stones which are found in beds in mountains, have been formed at the same time with the rock, and that those which are found in veins are of posterior formation to the rocks or strata of which the mountain is composed.

According to my observations, some of the metal formations are very old, and others have been formed

formed afterwards, and successively. *Tin* appears to me, if not the most ancient, at least one of the oldest metal formations; for I have never found it in floetz rocks, but sometimes in porphyry rocks. It is also one of the rarest metal formations. The ores of *molybdæna*, *tungsten*, and *wolfram*, appear to me to be nearly of the same antiquity; for they are seldom found, except in beds of tin; from which I infer that they have been formed nearly at the same epoch with that metal. The ores of *uran* and of *bismuth* appear to me to be of newer formation, although they have only been found in the primitive rocks, at least if they occur in transition or floetz rocks, it is unknown to me. *Gold* and *silver* appear to be of a still newer formation, and in some instances very new. These two metals are chiefly found in primitive rocks; sometimes, however, though very rarely, in rocks of a newer formation. *Mercury* is commonly found in primitive rocks, (with the exception of the oldest member of the series), and sometimes even in floetz rocks. The formations of this metal are very scarce, and appear to have been produced at very different epochs.

At *Rosenau* in Upper Hungary, at *Schönbach* in Bohemia, and at *Hartenstein* in Saxony, we meet with a *cinnabar* formation, which occurs in clay slate,

even

even in chlorite earth and talc. The cinnabar is united with quartz, calc spar, and sparry ironstone; also with copper and iron pyrites, iron-glance, and iron mica. This is probably the oldest of the cinnabar formations.

The formations of *copper*, of *lead*, and of *zinc*, are very numerous, and of very different ages. The *cobalt*, particularly *glance cobalt* and *copper nickel*, are of a very new formation. These metals are frequently found in transition and flöetz rocks, chiefly in *Mansfeld*, *Thuringia*, and *Hesse*. The white ore of cobalt, which is found at Tunnaberg, and Los in Sweden, and at Modum in Norway, as well as at other places, is of an old formation, for it is only found in beds in primitive rocks.

The grey antimony ore is of a middle age. I have never seen it in the transition or flöetz rocks, nor among the oldest metalliferous formations. *Arsenic pyrites* is an old production, but of various ages; for it is found in considerable quantity along with tin ore and lead glance, but in smaller quantity, with copper pyrites and arsenical silver ore. The *iron* formations are indisputably the most numerous, and of almost every age; yet, in attending to their age, we can point out different principal formations. The *magnetic* ironstone formation, which occurs in primitive mountains,

mountains, but more particularly in primitive limestone, seems to be the oldest formation of this metal. The *red* ironstone formation is a great deal newer; that of the *brown* and *sparry* ironstone is still newer; and the *clay* ironstone appears to be yet more recent. One of the newest is the *clay* and *magnetic* ironstone, which occur in rocks of the trap formation; and the newest of all is the *bog* iron ore. The formations of *iron pyrites* are almost innumerable, and are of every age, even to the latest period. The oldest veins are the only ones that do not contain some of it. The different *manganese* formations appear to be of an intermediate age. The oldest transition rocks are most productive of metallic ores: of this kind are grey wacke, transition limestone; to which we may add the oldest or first sandstone formation. The newest limestone rocks, and alluvial hills, are as unproductive of, as the oldest and intermediate ones are rich in, metals; and iron is the only metal which we find in them. It is only in some of the rocks of the coal formation that we meet with some insignificant traces of *lead* glance and *iron* pyrites. With respect to the relative age of some fossils of the other classes which compose the substance of veins, it appears to me that felspar, schorl, topaz, and even beryl, are found in the oldest veins. Veins, which contain,

in

in their substance, grey and green mica, also appear to me to be very ancient. All the calcareous fossils appear to be newer; and of these apatite and fluor spar are the oldest. Heavy spar is newer, and perhaps one of the newest substances which are found in veins. Perhaps quartz is the oldest of the veinstones, and sometimes it seems to be produced at all periods. The wacke and basalt which are found in veins are of very recent formation, and so are coal and rock salt. It is very worthy of remark, that, in the substance of the primitive rocks, we never meet with the smallest trace of inflammable or coaly matter: it is only in the oldest of the rocks of posterior formation that they begin to make their appearance. Rock salt appears to be of a very late formation.

The whole of this subject requires to be treated of more at length. What has been said on it in this paragraph, must only be considered as a slight sketch, from which it will appear how necessary an intimate acquaintance with the rocks is, to enable us to acquire a perfect and complete knowledge of the formation of veins.

§ 77.

I shall now produce a few particular examples, which will serve to give a more clear idea of these different

different metal formations, which have been formed successively, at periods more or less remote from one another. To be able to say something positive on this difficult matter, and which requires much skill on the part of the observer, it will be necessary to bear in mind what has already been said (§ 32.) on the principal and distinctive character of veins; viz. on the analogy or resemblance between the matters of which they are composed, (veinstones, ores, their species and varieties), and on the age of veins, as well as that of beds.

Native *gold*, among others, occurs in very different formations, and which have been produced at periods very distant from one another. What difference is there not between the native gold formation found in the beds of primitive rocks, in gneiss, clay slate, and mica slate, at *Ramingstein*, *Muerwinkel*, *Zillerthal*, and other places in the country of *Strasburg* and *Carniola*, and that formation found in veins in grey wacke rocks, in the transition mountains of Transylvania, at *Voeroespatak* and *Abrudbania*, and even in sandstone rocks at *Zalathna*. At the last place, native gold has also been found in semi-petrified wood, or rather in bituminated wood. The gold which is found at *Edswal* in Norway, at *Aedelsdorf* in Sweden, differs little in point of age from that found in the

the places mentioned above. The native gold found in veins in rocks of sienite and porphyry at *Schemnitz* and *Cremnitz*, holds a middle station between those kinds which have in this paragraph been cited as the oldest and newest formations. We may consider as the newest gold formation, that variety which is found in a micaceous and argillaceous sandstone at *Peresookoi*, not far from *Catharinenbourg*.

Native *silver*, as well as the ores of this metal, is of very various formations; but I have not been able to determine the relative age of the different silver formations with which we are acquainted.

The corneous silver ore appears to me to be the newest of the ores of this metal, for it is only found in the upper part of veins. At *Frankenberg* in Hessia, we find thin leaves of silver lying over petrifactions. In treating of the different depots of veins in the Freyberg district in Chapter X. I shall notice the different silver formations.

We are already acquainted with twenty different formations of *lead*. I have observed the following, and found them to be formations very distinct from each other.

1. *Lead glance* mixed with copper pyrites in quartz.

This

This occurs at Muerwinkel, and I believe at Hirzbach in the country of Salzbourg.

In the first of these places it is in beds.

2. *Lead glance*, with small granular brown blende, and slate spar.

It occurs at *Unverhofftes Glück* near *Beermansgrün*, not far from *Schwartzenberg*.

It is worked in the roof of a thick bed of limestone.

3. *Lead glance*, rich in silver, with small granular brown blende, and a little copper and iron pyrites in quartz.

It is found at *Hormersdorf* near *Geier*.

And is found in a bed of clay slate, approaching to the nature of chlorite slate.

4. *Lead glance*, rich in silver, with a large proportion of black blende, arsenic pyrites, and iron pyrites; sometimes with a small quantity of copper pyrites, more rarely sparry ironstone; in quartz, which is sometimes accompanied with a litttle brown spar.

This formation occurs in quantity in the mining district of Freyberg, and in other parts of the *Erzgebirge*, and almost always in veins.

In the 10th chapter, when treating of the mineral depot of Freyberg, I describe this formation as well as the 5th, 6th, 7th, 9th, 10th following.

5. *Lead glance*, very rich in silver, with black blende, a very small quantity of arsenic pyrites, and iron pyrites; in quartz and brown spar.

This formation occurs very often in the mining district of Freyberg, chiefly at *Brand*. It is found in veins.

6. *Lead glance*, rich in silver, with a little black blende, common iron pyrites, dark red silver ore, brittle silver ore, white silver ore, and plumose silver ore; in quartz, along with brown spar of a flesh red colour.

I have hitherto only seen this formation in the Freyberg district, chiefly in the neighbourhood of Brand, since which I have met with it at Kleinvoigtsberg; it occurs always in veins.

These two formations taken together, are described in the 10th chapter, and designated as the second depot of veins in the Freyberg district.

7. *Lead glance*, poor in silver, with a great deal of iron pyrites, black blende, and often ochry red ironstone; in quartz, with greenish clay more or less mixed with chlorite earth.

It occurs in many of the veins of Freyberg, chiefly in the district of *Halsbrücke*.

8. *Lead glance*, rich in silver, with yellow blende,

grey

grey copper ore, and common iron pyrites; in brown spar and quartz.

It is found at *Scharfenberg* near *Meissen*, and at *Kapnik* in Upper Hungary.

It occurs in veins.

9. *Lead glance*, poor in silver, with radiated iron pyrites, rarely with a little brown blende; in heavy spar, fluor spar, with sometimes a little calc-spar and quartz.

This occurs frequently in several places in the *Erzgebirge* in Saxony; in *Derbyshire* in England; near *Cimbrisham* and *Gislof*, in the province of *Schönen* in Sweden.

It is at all times found in veins.

10. *Common lead glance* and *compact lead glance*, with a little black blende, iron pyrites, and sparry ironstone.

It occurs in some of the mines in the district of Freyberg, in veins.

This appears one of the most recent formations of lead glance.

11. *Lead glance*, with a great deal of brown blende and sparry ironstone, a small proportion of iron pyrites, grey copper ore and a little copper pyrites; in quartz.

This formation is principally met with near *Strasberg*,

berg in the Hartz, and also, to the best of my knowledge, in the Upper Hartz, in the mines of *Kranich*, *Braune Lilie*, and *Zilla*. It also occurs in the valley of *Lahn*. In the Hartz it is found in veins; but in the valley of the Lahn, in beds.

12. *Lead glance* mixed with a large quantity of blende of a deep brown colour; in quartz.

This occurs at *Lauthenthal* in the Hartz.

13. *Lead glance*, with fine granular blende of a deep brown colour, iron and copper pyrites; in quartz.

It is found in the *Rammelsberg* at *Goslar*.

In a lying mass.

14. *Lead glance*, with copper pyrites, in calc-spar.

This is found in several of the mines of *Klausthal* and *Zellerfeld*.

In veins.

15. *Lead glance*, with a great deal of dark red silver ore, a small proportion of bright red silver ore, cobalt, and native arsenic; in calc spar.

It occurs frequently in the Hartz at *Andreasberg*.

In veins.

16. *Lead glance*, with calamine and a great deal of brown iron ochre.

It

It is found in great quantity in the flöetz mountains of *Cracow* in *Poland*, at *Tarnowitz* in Upper Silesia, and in the Ardennes.

At all times in beds.

This formation is the newest of all those we have described; it never occurs in veins, at least in mountains of considerable height; for it appears that the solution from which it was precipitated, had never attained a high level.

17. *Lead glance* in very small quantity, very poor in silver, partly disseminated, and partly occurring as a coating, with copper pyrites; in calc spar.

It sometimes occurs in veins, in the coal formation, and in thin laminæ in rents of the coal itself.

It is found in different parts of the coal mines near *Dresden*.

This formation is very insignificant in regard to its quantity, but very remarkable on account of its newness. It has already been noticed in the preceding paragraph.

Of these different formations of lead glance, there are five which occur in the Hartz; but it is very difficult to judge exactly of these formations, when a person has not himself seen the repositories where they

they are found, which is my case with respect to the formations of the Hartz; for it is not sufficient to have seen specimens, to be able to judge properly; we must have examined exactly and frequently the repositories or places where they occur; we must have ascertained the order and connection of the fossils which belong to the same formation; and we must have observed the varieties of the formation in different places, as well as the mutual relations of the repositories in which they are contained. It is only by correct and repeated observations, and by comparisons made with care, that we can ascertain that two formations which appear to bear a strong resemblance, are not to be regarded as one and the same, on account of essential differences which they present; that the varieties which the same formation shows, must make it be divided into several others; and that several others which are found together in the same repository, are to be considered as belonging to one and the same principal formation. In this way it may happen, that what I have said on the formations of the Hartz, may stand in need of corrections, or may receive many additions. Perhaps by repeated observations made in different countries, even in those at the greatest distance from each other, a greater number of lead glance formations may be discovered.

§ 78.

The researches which we have just mentioned in regard to different particular formations of metals or ores, lead to an observation not of less importance, viz. what are the ores and fossils which commonly are found together, and what are those which seem in some measure to exclude others. I shall only mention but a few examples of these.

Lead glance and blende, or at least *calamine*, always occur together. Lead glance is often also accompanied with copper pyrites. *Cobalt*, *copper*, *nickel*, and *native bismuth*, are commonly found together; yet native bismuth does not occur in the newer formations of the two former metals. *Tin* often occurs along with wolfram, tungsten, molybdæna and arsenic pyrites; likewise with topaz, fluor spar, apatite, schorl, mica, chlorite and lithomarge. *Brown ironstone* is generally found with sparry ironstone of a deep brown colour, black ironstone, manganese and heavy spar.

As to the fossils which seem mutually to exclude one another, I have to remark, that in those places where *tin* occurs, we rarely meet with the ores of silver,

silver, lead and cobalt; or with heavy spar, calc spar, and gyps. It is not common to find blende with copper pyrites. *Cinnabar* and the ores of *mercury* scarcely ever occur along with the ores of other metals, with the exception of iron ochre and iron pyrites. I know but of four exceptions to this general rule; viz. at *Moersfeld*, where the cinnabar is mixed with a little lead glance; at *Moschel Landsberg*, where it is joined with a small proportion of native silver, grey copper ore, malachite and copper azure; at *Schemnitz*, where it is found with lead glance, black blende, and copper pyrites; and at *Rosenau* in Upper Hungary, where it is joined with copper pyrites, sparry ironstone and iron mica.

With respect to *grey manganese ore*, I have seen it joined either with red ironstone, or with black ironstone and heavy spar.

The peculiarities which have been mentioned in this paragraph, may arise, in part, from the proportion of the constituent parts, and partly from a greater or less difference in the time of the formation, either of those fossils which are excluded, or of those which occur together.

§ 79.

§ 79.

It has been long observed, that certain ores and fossils, which form the substance of veins, are only to be found in rocks of a particular kind, and consequently that these fossils and ores do not occur in the other kinds of rock: thus *calamine* is never found in granite, gneiss, mica slate, or the other primitive rocks; in like manner, we very rarely see *silver* in granite rocks. The cause of this has hitherto been attributed to the immediate influence of the rock; but more accurate observation has shown not only that there is no foundation for this, but has even contradicted it, as shall be shown at § 90. Observation, and the new theory of veins, teach us, that certain metals and fossils are of very old formation, and that as the solutions from which they were produced, did not exist at the period of the formation of the newer rocks, of consequence they could not fill the rents that took place in them. Certain fossils are also of newer formation; and as the solutions which furnished them stood at a low level, and did not cover the older rocks, their precipitates could not enter into the fissures in them. It might also happen, that, though certain mineral substances were suspended in the general solution, that another formation could not take place, because the fissures in the rocks were already filled up.

§ 80.

A long period of time unquestionably elapsed before the principal formations of certain ores and fossils were completed: during this period, not only the already existing fissures were filled up, but even new rents took place, and were filled up with the same or nearly the same solution. In consequence of this, we find several particular formations of veins, which appear like so many subdivisions of the principal one: these veins, in traversing those that are older, though of the same principal formation, discover to us a newer formation. All these veins of a particular formation, belonging, however, to a principal formation, *always traverse all the veins of the principal and older formation, and in their turn are traversed by all the veins of a principal formation which happens to be of newer date.* I have remarked this to be the case with the veins of the Freyberg district, as well as of other places.

§ 81.

§ 81.

Before concluding, I must observe, that the substance of veins already existing, may, in process of time, have experienced certain changes and alterations; and this will be found to have actually happened in many places. The substance of a vein (or only one of the fossils which compose it), may have been entirely decomposed, or its composition may have been changed by the addition or subtraction of one of its constituent parts; and one fossil may in this way have been transformed into another. A similar change may have been produced either by a solution which may have been introduced into a vein through an opening that existed in it; or by the action of the atmosphere which penetrated to it; or by water which had filtrated through the earth and rock composing the mountain. In the first of these cases, it is a difficult matter to ascertain and distinguish the change produced by a new formation which has been joined and intermixed with an older. In the latter case, (viz. of infiltration), which could only happen in the vein, to a small depth below the surface of the ground, it may have been the cause of a new production of ores.

After

After the remarks which have been made, it will be supposed that such transmutations and transformations ought to be very rare, which is really the case; they only produce ochres and metallic oxides, very rarely native metals (copper perhaps alone excepted), never ores properly so called. Among earthy fossils, fluor spar seems susceptible of such a decomposition, which it seems sometimes to have undergone. The stalactites which are sometimes found in mines, are proofs of the decomposition of calc spar, brown spar, and of some ores.

CHAP.

CHAP. VIII.

A SHORT REFUTATION OF THE ANCIENT THEORIES OF THE FORMATION OF VEINS.

§ 82.

In the second chapter of this treatise, I have given an historic sketch of the ancient theories of the formation of veins, without adverting to the objections which might be brought against them; conceiving that they would fall of themselves, and be completely refuted by the exposition and proofs of my new theory, as well as by my answers to the objections which have been, or may be, urged against it: in this way I have avoided much delay. But as there are several things to be said against these old hypotheses that I have not had occason to notice, which evince and place in a clear light the falsity of these theories, I shall subject them to a new examination and

and consideration. All the hypotheses which have been treated of in the second chapter, are either wholely or in part erroneous, and may be comprised under four principal theories.

(*a.*) Veins have been produced at the same time as the terrestrial globe (§ 7. 11.).

(*b.*) Veins, and their branchings, are to be considered as the branches and twigs of an immense trunk, which exists in the interior of the globe (§ 16.).

(*c.*) Veins were fissures which have since been filled up by degrees with mineral matters (§ 7. 9. 13. 15. 17. 18. 19. 20. 22. 23. 25.).

(*d.*) The matter of veins is neither more nor less than the substance of the rocks, which nature has changed and transmuted by the action of certain solvents which have been introduced into small fissures (§ 14. 21. 24.).

With respect to the old opinion, which attributes to the sun and planets an influence on veins, this, as well as the virgula divinatoria, belongs to the mystic astrology of ancient times, and the system of effluvia: it is so inconsistent, so incoherent, so superstitious, and has so long since been committed to oblivion, that it does not deserve the name of a theo-

ry,

ry, far less a refutation: indeed it hardly ought to be named.

§ 83.

The first opinion, viz. that which ascribes to veins an origin as ancient as that of the rocks in which they occur, or that they were formed at the same time, was styled by Agricola the opinion of the vulgar; but it has been since brought forward and supported by Stahl and Juncker. Its refutation is rendered complete by the following observation; viz. that in the mass of rocks that are evidently of new formation, we find veins which present the same peculiarities and phenomena that others do: for example, we find true veins in rocks of the coal formation, in those rocks containing beds of bituminated wood, and, in general, in those rocks which not only, by the numerous petrifactions of the vegetable and animal kingdoms which they contain, but by their stratification, are evidently shown to be of new formation, and to have been formed by a succession of precipitates, produced from a vast solution, and arranged above one another. If the mountains which contain these veins were not formed until a considerable time after the existence of our globe, the veins which are contained in them must, of necessity, be much

much less ancient than the rocks which they traverse. But further, if, as all observers agree, veins which occur in the primitive rocks have the same peculiarities of structure and position as those that are found in rocks of newer formation; it is evident that both the one and the other must have the same origin. Finally, if the stratification and the nature of the rocks of the most ancient mountains exhibit an origin similar to that of the newer formations; these primitive mountains must have been formed successively and by degrees, and the veins which they contain must consequently be of still newer formation; so that these veins, and, for a stronger reason, those also which occur in the newest rocks, must have been formed after the creation of the globe, during the term of its existence, and during those revolutions which it has successively undergone.

This opinion, which ascribes to veins an origin as ancient as that of the globe, is also contradicted by the observations which show that veins are of different ages with respect to one another, as well as with respect to the rocks in which they are enclosed. It is also known that veins contain substances which evidently show a recent origin; these are rounded masses or rolled pieces, fragments of the rock or substance of the veins, and petrifactions. Farther,

we

we know, that new rents are formed from time to time; that they are formed even in our time; that these rents are by degrees filled up; and that there are formed in them substances perfectly the same as those contained in other veins.*

§ 84.

The second hypothesis, that of Lehman, which considers veins as the branches and twigs of an immense trunk of mineral matter contained in the interior of the globe, of consequence ascribes the ori-

X gin

---

* I find that Dr John Junker treats of the formation of ores and metalliferous veins, in the second chapter of his History of the Old Theory of Veins, of which I omitted to make mention in the proper place. This zealous follower of Becher and Stahl treats of this subject in the second part of his complete Treatise of Chemistry, (translated from the Latin into German, 4to. Halle 1750), and very particularly in the 32d section, which is written on metals in general. This work ought to be read by all who are desirous of studying the literary history of veins, and of becoming acquainted with the singular opinions which then prevailed, not only on this subject, but on others of a different nature.

Junker was a great collector of facts in chemistry in general, as well as on this subject in particular.

gin of veins to a sort of vegetation or organic movement. Such an hypothesis may be the effect of the dreams of a man who has never beheld the interior structure of a mountain, but can never be admitted by miners and experienced geognosts; for they know, that veins, in proportion as they sink deep, become narrow, and at last terminate in wedges (§ 41.). In truth, we have, as yet, only reached the lower extremity of small veins, or of such as are of moderate size; and if we have not yet found the extremities of large veins, it is by reason of their great size, and that they sink to a depth greater than what our labours have yet been able to follow. But the perfect resemblance which in every respect we see between the larger veins, and those of less magnitude, permits us to conclude, by analogy, that the former terminate precisely in the same manner as the latter: what ought to confirm us in this opinion, is, that most of the large veins become narrower at a great depth. This contradicts entirely the pretended communication supposed by Lehman to exist between veins and the interior of the globe.

It is, besides, impossible to conceive that a kind of vegetation can take place, which shall traverse the solid and compact substance of a rock. Nature no-

where

where presents any fact which has even the most distant resemblance to this phenomenon: on the contrary, what we see in veins is a perfect contradiction to this hypothesis. To conclude, we may urge against it what has been said at the end of the last paragraph on the subject of rounded masses, the fragments of rocks, petrifactions, &c. found in veins.

This opinion of Lehman on the formation of veins will accord very well with that of some modern philosophers, who regard our globe as an organized being (an animal of an extraordinary size).

§ 85.

The third hypothesis, which has more the form of a theory than that I have just refuted, has many partisans: it contains two principal propositions.

1. The spaces which veins occupy, are nothing else but fissures which have been produced in the interior of rocks.

2. These fissures have been afterwards filled up with the substance of the vein.

I not only admit the first of these propositions, but I think the proofs which have been given put the matter beyond the possibility of a doubt. I think (in § 29. 39. 40. 53. 57.), I have assigned, in a more particular and exact manner than any one, the causes

causes which have produced these rents, (most authors ascribe their origin to drying and earthquakes).

The second proposition, taken generally, and in the manner it is here expressed, accords perfectly with my theory; but the opinions differ as to the mode in which the rents have been filled: these differences of opinion constitute as many different branches or subdivisions of the theory; they all differ very essentially from mine, and I consider them all equally erroneous and inaccurate. The following are the subdivisions of this third theory.

(*a.*) The spaces which veins occupy have been filled during the deluge (mentioned by Moses), with earthy and stony matter.

(*b.*) The substance of veins has been formed by mineral substances which have been carried along by the waters, which, after having traversed the rocks, have entered the fissures, and there deposited these substances; the exhalations arising from the bosom of the earth, acting upon such of these materials as were fit for such a transmutation, have converted them into ores.

(*c.*) The substance of veins has been formed by depositions and precipitates produced by the waters, which, having penetrated to the heart of the rock, there

there dissolved and carried along with them into the veins, the substances of veinstones and ores which are now found existing in rocks.

I shall now state the objections which may be urged against each of these theories individually.

§ 86.

The first opinion was embraced by Stahl, but has been since abandoned by almost all the world (§ 11.). According to it, the substance of veins is to be regarded as a precipitate arising from a general inundation, which has been but of short duration, such as the universal deluge. This opinion falls of itself, when we take into consideration not only the internal structure of veins, which shews them to have been formed by degrees, and to have arisen from a chemical solution, but also the marked variety of their substance; this variety, joined to the other peculiarities which veins exhibit, proves them to have been formed at different epochs, very remote from one another. It may be stated as a farther objection to this hypothesis, that in the primitive mountains we see a great number of veins filled with fragments of rocks, which in other places constitute mountains of newer formation; a circumstance which shows that the continuance of the production of veins has been as long

as

as that of the formation of rocks themselves. From hence it follows, not only that the duration of the deluge was a great deal too short for forming the substances of veins, and that it is a very trifling period compared with what was requisite; but also that veins would neither present that uniformity, nor the structure which we observe in them, if they had been formed from a mechanical solution, which must be supposed in admitting the hypothesis of Stahl.

§ 87.

The second subdivision of the third theory, is the opinion of those who consider the metallic matters found in veins, to have been produced by subterranean exhalations exerting an influence on the earthy and stony substances already existing in veins, and which were fitted for this transmutation. This hypothesis, first proposed by Becher (§ 10.), and adopted by Henckel and several others, is combated by the following facts. We do not find in any of our mines the least trace of any subterranean exhalation. The position and form of ores, found in veins, show them to have had the same origin as the veinstones. The choke damps, formerly considered as metallic or arsenical exhalations, are now ascertained to be, at

least

least in great part, nothing more than carbonic acid gas, and the causes of it are now well known. The regularity of order and position, which ores observe in regard to the veinstones, is so well known, that I shall content myself with observing, that I have often seen in veins druses lined with crystals of a large size, whose superior part was covered with ores and veinstones lying over one another: thus, their position evidently shows them to be depositions coming from above; and that it is impossible that this phenomenon could have been produced by exhalations.

In the cabinet of minerals belonging to the academy of Freyberg, there is a very fine and highly interesting druse of calc-spar, formed by pyramids of calc-spar, on one of the sides of which are alternate layers of quartz, lead glance, and sparry ironstone.

The ores formed in the beds of mountains exclude entirely all idea of their formation by means of exhalations or a kind of sublimation. These ores are nevertheless absolutely the same with those found in veins; they are arranged and disposed in the same manner in beds and veins, which denotes them to have a similar origin in both cases: now it is impossible to conceive how these ores could have been produced

duced in the beds by exhalations. In § 90. will be found two farther objections against this hypothesis.

§ 88.

I shall, in § 90. refute the system of those who consider the substance of veins as a precipitate formed from the waters, which had penetrated through the rocks into those spaces which are now occupied by veins. I confine myself to that chemical property ascribed to certain earthy or stony substances composing veins, a property which is considered as necessary to the formation of metals, and, in virtue of which, these substances receive and fix the pretended generative exhalations, so that they serve both as a chemical base, and constituent part, in the formation of metals; in short, they are what have been called the matrices of metals. This chemical affinity between earths or stones, and the metals found in the same vein, is nowhere confirmed either by positive and convincing observations, or by experiments made in our elaboratories; on the contrary, we remark a perfect uniformity of position and form in the metallic substances which are found in the same vein. No conclusion can be drawn from such and such metals being found with such and such stones.

We

We might with as much reality infer, that the metal is the matrix of the stone, as it was formerly said that the stone was the matrix of the metal. This ancient doctrine of the matrices of metals is an additional proof, that when an opinion is once adopted, a person thinks he can observe, in every part of it, proofs, which not only have no relation to it, but even prove the contrary. Such modes of observing are singularly prejudicial to the advancement of science, and we unfortunately see but too many instances of the kind now-a-days. All this theory of the matrices of metals (which has been the cause of regarding veins as produced by the transmutation of the substance of the rock), rests upon two pretended propositions in chemistry, which are false, but which were formerly generally received: viz. that metals are composed of earths, partly not metallic; and that from simple earths, not metallic, metals may be artificially formed. This last proposition was founded on an experiment of Becher's, as celebrated as it is insignificant, by which a small quantity of iron was obtained by working and distilling lintseed oil and mastich with pounded bricks.

§ 89.

On the subject of fermentations, which some authors pretend to be continually going on in the interior of rocks, (an opinion first delivered by Henkel, as far as I know), I cannot help remarking, that we do not see the smallest trace of such effects or phenomena in the interior of rocks, nor even in mineral repositories, provided these repositories are shut and have no communication with the atmospheric air. Should the efflorescence and decomposition of pyrites be taken for such a fermentation, which, however, appears to me to be very different, I would reply, that such a decomposition is peculiar to iron pyrites alone; for the efflorescence which is sometimes observed in other varieties of pyrites, only happens in those that contain a proportion of martial pyrites. 2. All the kinds of pyrites are not liable to this efflorescence; only the hepatic, radiated, and such others as come near to these varieties. 3. Every kind of pyrites, even that which is most susceptible of this efflorescence, does not undergo it in close places, such as mineral repositories commonly are, but only when exposed to the contact of atmospheric air, and chiefly to the alternation of dryness and humidity,

humidity, cold and heat. The heaps of rubbish brought out of the mines (Halden) containing pyrites, emit a strong smell when exposed in the open air to the rays of the sun when very warm: if the decomposition was effected as easily in the interior of rocks, would not our organs of smelling be affected in a similar manner in close places, such as mineral repositories; this however never happens, or at least very rarely even in veins which abound with pyrites.

I have likewise found, that what has been said of the great heat observable in mines of pyrites is totally void of foundation. In winter, the air in mines is, without doubt, warmer than the atmosphere; it may, however, happen that some mines and galleries, where there is not a free circulation of air, may be warmed more than others by the workmen and the heat of their lamps. But with regard to mines of pyrites, of which there are great numbers at Freyberg, I have never remarked their temperature to be higher than that of others; even in the pyrites mines of *Catharina* at *Raschau*, and *Stam-asser* at *Graul*, I was not able to observe the heat which had been mentioned to me, and which I was anxious to discover.

Thirty years experience and observation have fully convinced

convinced me of the little reliance to be placed on the *signs* of the existence of metallic veins: according to the ancients, these were,

(*a.*) Igneous meteors, seen in those places where veins crop out.

(*b.*) The celerity with which snow melts.

(*c.*) The diminutive size and sickly appearance of plants and trees growing on mountains which contain metals.

Our ancestors were much inclined to believe things, which we now-a-days do not credit, and scarcely ever speak of. Such pretended appearances have yielded to real knowledge. We must excuse them, for having believed and admitted things (respecting the formation of veins and minerals), which tallied with the theories of their time. They wished to appear to their cotemporaries as persons possessing experience and knowledge. If this was neither candid nor proper for philosophers, it shows at least that they valued the theory and experience of which they voluntarily made a display.

§ 90.

A great many naturalists consider the substance of veins, (ores as well as veinstones), to have been formed, or rather deposited from water, which, passing

sing through the substance of rocks, dissolved certain earthy and metallic substances with which they became charged, and carried them along with them into the cavities now filled by veins where they were deposited. To this theory, I object, 1. That the substance of veins (chemically speaking), is generally of a nature very different from that of the rock in which they occur. 2. That in one and the same mountain we find veins, which, though interwoven, are of very different nature. 3. On the contrary, it often happens that veins of the same nature are found in rocks of a very different kind. For example, in our mountains of gneiss, among the metalliferous veins, we often find veins of brown spar or heavy spar. How great is the difference between the nature of these minerals and gneiss? and for a stronger reason, what a difference is there not between gneiss and metallic ores! Farther, in Derbyshire, all the veins which occur in the limestone rocks of newer formation, consist of heavy spar and fluor spar; in the limestone rocks of *Saalfeld*, heavy spar is the only kind of veinstone which occurs. We will hardly meet with any place where we shall find, in so remarkable a manner as in the Freyberg district, an assemblage of veins composed of several different materials in one kind of rock, and even in the same

kind

kind of country. We there find, lying near to one another, veins of quartz, of heavy spar, of heavy and fluor spars, of loam, and these veins often crossing. We will see, in the tenth chapter, that the ores contained in these veins are almost as various in their nature as the veinstones. How is it possible that one and the same mountain, one and the same kind of rock, (for in almost the whole country we see nothing but gneiss), could have furnished such a variety of different formations: these formations are indeed contained in veins that have different directions, but they mostly all join together, and must of consequence have been filled by the same canals (of infiltration).

When in one and the same mountain, in one and the same kind of rock, three veins cross, one of which contains heavy spar, the other quartz, and the third brown spar, (which is by no means a rare occurrence in this district); how, in this case, I would ask, where these three veins have the same canals, is it possible, that water could have taken up heavy spar to be deposited in one vein, that in the same rock it should only have been impregnated with quartz to be deposited in the second vein, and and brown spar for the third. At *Ehrenfriedersdorf*, in the same place, there are veins of silver and of tin,

tin, which have a different direction, and cross; how, I ask, in the same rock can water be loaded with particles of silver to be carried into one vein, and with particles of tin to be conveyed into another? Is it possible to conceive that this really can happen?

With respect to the existence *of one and the same formation of veins* in rocks of different natures, I have to observe, that the oldest lead glance formation in this country (which consists of lead glance, black blende, arsenic pyrites, copper pyrites, iron pyrites, quartz, and a small quantity of brown spar), occurs,

At Freyberg, in rocks of *gneiss*,

Near Mohorn, in rocks of *porphyry*, and

Near Mintzig, in rocks of *clay slate*.

In like manner, the lead glance formation, poor in silver, and containing heavy spar and fluor spar occurs,

Near Freyberg, in rocks of *gneiss*, and

At Derbyshire in England, in *limestone* rocks.

This circumstance, of the existence of a vein formation in rocks that are, chemically speaking, of different natures, evidently shows that the rocks of a mountain have no influence upon these formations.

The falsity of the theory which we have just been attacking

attacking in this paragraph, is still farther demonstrated by the internal structure of veins. Veins, as has already been said, are composed of layers parallel to their sides; crystals, and their impressions, show clearly that the external layers have been first formed: now, had veins been formed by internal infiltrations, these first layers must have shut up the openings of these pretended canals; in this way the entrance of all new matter must have been obstructed, and the formation of the vein stopt.

We may still farther observe, that the circulation of the waters, which this hypothesis necessarily requires, could never take place in those parts of veins and their branches which are below the level of the foot of mountains. Farther, in the open rents and fissures of veins found in mines, there often is a great deal of moisture; but that in a compact rock, which is little rent, chiefly when the strata are horizontal, the rock is often very dry, to the great prejudice of the miners, who are driving galleries across them. Again, the waters, which having penetrated into the rocks, dissolved and carried with them a sufficient quantity of earthy and stony matter necessary for the formation of such a number of veins as are found close together at Freyberg, or of the large veins of the Hartz, ought to have left considerable open empty

spaces

spaces in the rocks, which, however, is nowhere to be seen. Besides, in those parts of the rocks at a distance from veins and metallic beds, we do not find the least indication of metallic substance, if we except a few insignificant particles of iron. Finally, we may object to this theory, that if the waters which have filtrated through the rocks have dissolved a great quantity of metallic matter, and become charged with it, the water which flows from metallic veins ought to carry along with it a quantity of metallic particles, which, however, does not occur even in those countries which contain the greatest number of mines: such water rarely contains a small proportion of iron; scarce ever any particles of copper; never silver, lead, tin, zinc, cobalt, mercury, arsenic, &c.

The observations which, in this paragraph, have been made on certain districts, in which we find veins near one another, each containing ores of different natures, as well as on the obstruction to the canals of infiltration by the first layers of veins formed immediately on their walls, may be equally urged as objections against the theory refuted at § 87, which theory ascribes the formation of veins and ores to metallic exhalations arising from the interior of the globe.

§ 91.

It remains for me now, to examine and refute the fourth theory of the formation of veins. Zimmerman, who is the author of this, supposes the rock to be traversed by small rents and fissures, into which certain solvents have been introduced that have transformed the matter of the rock into that of the vein. The objection to be made to this theory, and which is sufficiently strong to overturn it altogether, is, that not only all correct observations made on the substance of rocks and veins, (as well upon the fossils in particular, as upon the arrangement and disposition which they have among themselves), but also all experiments hitherto made in the laboratories of the chemists, absolutely contradict the transmutation of the simple and elementary particles of bodies, such as the transformation of one earthy base into another, or of an earth not of a metallic nature into a metal, &c.

The intransmutability of the simple and elementary parts of bodies is one of the fundamental and immoveable pillars of chemistry.

We nowhere see even a trace of such transmutation; and hitherto not the least fact, or one single experiment,

experiment, can be quoted really in proof of it. For when one mineral is changed into another, that is to say, when a mineral loses one of its constituent parts, and acquires a new one; for example, when native silver is changed into silver glance, when felspar is changed into porcelain earth, or when iron pyrites becomes iron ochre; then one mineral is actually changed into another, but the simple elementary particles have not suffered any change in themselves; it is always the same argillaceous earth, the same silex, the same iron, the same silver. Even when, in the same specimen, we meet with particles of heavy spar and quartz, or siliceous and calcareous particles mixed together, we cannot conclude that the spar has been changed into quartz, or the siliceous into the calcareous earth, and reciprocally: but rather, that the stones which compose the same specimen, have been formed beside one another at the same or different times. In § 58, 59, 70, 71, 72, 73, I have noticed several of these pretended proofs of the transmutation of the rock into the substance of veins; and I have shown how unconvincing, and even how inadmissible, these proofs are.

I shall, however, afford the followers of this theory an opportunity of placing it, if it be possible, on a more

more solid foundation, and of bringing forward clear proofs, such as are capable of supporting it, which they will have it in their power to do by answering the following questions. What is, I ask, in the first place, this active chemical agent possessing such powers, which can change, and must have changed, quartz and mica into gold and silver, and sandstone into coal? What is the nature of this powerful agent? Do we know any thing similar to it in nature, or which approaches near to it? Has there ever been observed any where such a matter which still continues to act? Is there one certain experiment, should it be but a solitary instance, which proves that this substance, capable of producing such great changes, has, in spite of all chemical theories and experiments, converted any mineral into another, whose elementary parts or base only are different? In such a fact, which must, however, be brought forward to prove this theory, who can certify that the mineral said to be transformed was formerly and certainly of a different nature? Who can say of any mineral that we see, that it was not precipitated from a solution, but that it was formerly another mineral, which by the assistance of some known or unknown principle has been transformed, and has become such as we now see it? Answers given to these questions, and deli-

vered

vered categorically, will be proofs of the truth and validity of this theory; but if they be not given, or cannot be made, then is this theory absolutely divested of every kind of support.

As to what remains, the phenomena and peculiarities which veins present are any thing but explained by this theory; it cannot account for even the most simple facts, because it is at variance with the whole of them. How will it account for the internal structure of veins, when they are composed of different layers of crystals placed upon one another, in the centre of which open druses are contained? We see clearly, that it is by one crystallization that the two outer layers are formed on the walls and roof; a second crystallization has produced the two layers next to the first; that part of these new layers which has been applied to the former, has received the impression of its crystals: on these second layers, on each side, newer ones have been deposited, and so on in succession, till we arrive at the middle part of the vein, which often presents open druses. How is it possible to explain all these peculiarities by the theory of transmutation? Is it possible to conceive that the crystals, often of a great size, which are seen in the heart of a vein, can have been produced by the transmutation of a little bit of gneiss, or of granite? How

can

can such a transmutation of the rock produce the open druses which we see often of very considerable size in veins?

Let us pass to the substance of the vein, and I ask, how can this theory account for the great diversity which we observe in the substance of veins, and the difference in the different kinds of ores? Will it explain how certain veins have been filled only with coal or rock salt. Farther, how does it happen, that, in the same country, we find veins so different in their substance, (for example, veins of lead glance, veins of silver, and veins of copper, &c.), each of which has a different direction? There are veins whose substance is very compound, in which one single formation contains sometimes six or eight kinds of different fossils, in which we sometimes find two or three formations together. I ask if the transmutation of the rock has produced all these different fossils? One single principle or agent cannot surely have changed gneiss into lead glance, silver, pyrites, quartz, heavy spar; and if to change one rock into different kinds of fossils, or to transform different kinds of rock into the same fossil, (such as to convert into lead glance, either a small piece of granite or limestone), if, for either of these transformations, a particular agent were necessary, where should we

be

be with such a variety of agents fit to produce all these transmutations? I ask finally, if the theory of transmutation can account for the rolled pieces, the petrifactions, the fragments of rocks which are found scattered through the substance of veins? The presence of such in veins banishes all idea, even the most remote, of transmutation.

These are not yet all the objections and difficulties which may be opposed to the admission of this theory. If veins owe their origin to a substance introduced into narrow fissures, and which has there produced a transmutation of the substance of the rock, no reason can be assigned why this transmutative substance should not equally have insinuated itself into all the fissures of mountains, particularly into the narrow fissures (Schichtungs-klüfte, fissures of stratification), which separate the beds as well as the strata of rocks from one another, and why it should not have transformed the whole rock into veinstones and ores. Farther, it is difficult to conceive, how the capillary fissures, which this theory admits, could extend above a mile in length, and to a much greater depth; such a supposition is, however, necessary to explain the formation of large veins. No more can we see why such transmutation should not happen even in our own time. The followers of the theory of transmutation have

have not yet determined whether they admit the formation of rents and fissures in rocks from time to time, nor whether they can admit these fissures to have been filled from above by matter which has been precipitated in them. If they refuse to admit the formation and existence of rents, I shall remit them to § 40. where I have shown that they continue to be formed in our own times. In admitting these rents, they must allow that there are veins which have originally been fissures, but have since been filled up with matter precipitated in them. Then I would ask, in what respect do veins produced in this manner differ from those formed by transmutation, and wherein does this difference consist? If there be none, then is the theory of transmutation useless. Lastly, We may ask, in what manner have minerals, such as we find in veins, been produced in the beds of mountains? It is not possible to suppose the transmutation of beds. For if, on the surface of the globe, as well as in the interior, the whole had experienced a transmutation; which part of its substance first underwent the change, and became what we now find it to be? But if the ores and minerals in general which we find in the beds of mountains have been formed by precipitation (in the humid way), what hinders us from supposing the same substances

stances when they occur in veins to have been formed in a similar way? And granting this, the theory of transmutation becomes superfluous.

I hope what I have said on the fourth principal theory of the formation of veins, is enough to prove its insufficiency, and to show how inadmissible it is.

CHAP. IX.

OF THE APPLICATION OF THE NEW THEORY OF VEINS TO THE WORKING OF MINES.

§ 92.

AFTER having completely finished the exposition of my new theory, after having answered the objections which have been and may be brought against it, and after having refuted the former hypotheses; it only remains for me to give the application of this theory to the art of working mines, and so to show its utility; an undertaking which I enter upon the more cheerfully, as I have studied this subject not only as a geognost, but as a miner. The advantages to be derived from this theory are very numerous, yet I shall state them in a few words.

§ 93.

The first important advantage which the miner derives from this theory is, that it makes him acquainted in a more perfect manner with the physical state of the different districts and ranges of mines, and thereby serves as a guide to direct him in the examinations which he is obliged to make in the whole of a district in general, and particularly in searching for those places that are most convenient for working.

The new theory teaches,

1. To know the different depots and formations of ores which occur in a certain district, and

2. To determine their extent and natural limits; and,

3. It directs and fixes the attention on the crossings of the different depots.

§ 94.

Secondly, It enables us to distinguish the *particular repositories*, which under the form of precipitates, *cover the bottom of the reservoir* in which the great solution from which they were formed was contained; such as the *beds* and *lying masses* of rocks, together with

with those, which, under the form of precipitates, were formed in deep openings; such as *veins* of every kind, *stockwerkes* and *standing masses*. As each of these repositories has its own mode of formation, different from that of others, the new theory enables us to direct the labour of working the mine in a manner conformable to the nature of the substance worked upon.

§ 95.

It enables us, in the third place, to determine what veins ought to be worked in preference to others. For although this may be determined

By the kind and richness of the ore which the vein chiefly contains;

By the size of the vein;

By the quantity of the masses of ore which it contains, and their distance from one another;

By the abundance of metal which they contain, and,

By different other circumstances which favour or retard the labour of working—

Yet we ought besides to have an eye to the number of precipitations of ore which may have been produced in the vein; that is, to the number of metallic formations which it may contain.

§ 96.

§ 96.

This theory serves, in the fourth place, to facilitate and complete the investigation of the parts of the vein where the ore is found. For when I know where, and in what manner, a metalliferous formation occurs in a vein, I am able to go more directly to the mass, or the place where the ore is contained.

§ 97.

Fifthly, It enables us, principally by being able to determine the relative age of veins, to judge with certainty of the peculiarities which they present when they meet, traverse, join, ramify and derange or intersect one another, which may assist the miner in extricating himself from the difficulties into which he may have been thrown by these accidents.

Farther, my theory throws light on the relation of veins to the beds and strata of mountains in which they occur, and reciprocally on the relation of the beds and strata to the veins.

§ 98.

When we meet with a vein, we can by means of this theory, better than by any other, determine whether it be one that we are already acquainted with, or whether it be a different one. I believe that in ascertaining this point, we have hitherto proceeded too much according to rule, without having paid sufficient attention to nature.

§ 99.

In the seventh place, this theory teaches us whether or not the rock contributes to the formation of metallic veins. It shows us that the rock, considered in itself, has no influence on the quality of the ores which are found in the veins, but that their formation depends upon the age, or rather the existence and disposition of the mountain at the time it was covered by the solution from which the matter of the veins has been precipitated. It explains to us how new rocks have, in forming, covered the more ancient ones; so that the veins which these last contain, not being able to traverse the rocks of posterior formation, canno show themselves.

§ 100.

§ 100.

Eighthly, It furnishes a method and language, as well for observation and practical investigations and operations, as for the theoretic and scientific exposition of this part of mineralogy and the science of mines.

In continuing to investigate and apply this theory, new advantages will very likely be discovered, for it is as yet but in its infancy.

§ 101.

In the working of mines, cases may occur to which my theory may be immediately applied; more frequently, however, but little advantage can be expected from a solitary application made without regard to the system itself in all its bearings; it may sometimes even lead into error, and such applications may beget distrust in it. For with the exception of a few general propositions which are easily understood, it is too complicated, and too much above the attainments of the generality of persons employed in the service of the mines. It requires a degree of combination and sagacity in the observer, who ought to be familiar with the theoretical principles, accustom-

ed

ed to make particular observations, and in the habit of combining them. Persons who possess these qualities, can alone make use of my theory with profit; so the application of it can chiefly be made by directors and counsellors, who superintend the works of a whole district: it may be useful to them if it be supported by sufficient observations, if it be in a manner made local, and if their observations be collected and arranged in a proper manner. Thus, in all the labours of working that may be undertaken, it will afford light to these directors and counsellors, as well as instructions on the state and nature of the mineral repositories, but more particularly the veins which are found in their districts.

§ 102.

If a person wish to apply, with advantage, my theory to the working of the mines in a district, it is absolutely necessary to commence by collecting and arranging in a proper order all the knowledge which he possesses of the repositories of the fossils known to be in the district, and capable of being worked, as well as ascertaining the relations which these repositories bear to one another. This may, and ought to be done in two different ways; in the first place, a geognostic plan of the district ought to be constructed

structed according to the principles of the new theory, and there should be joined to it a geognostic description of the district. This plan and description should be constructed according to the same principles, be equally complete, and have such a relation to each other that the same explanation may apply to both. If these two works be properly completed and made with accuracy, they will form a groundwork, on which may be constructed in the surest manner, all the plans and schemes relative to the working of mines in the district where they occur.

§ 103.

The geognostic plan of a country should consist of two maps or principal designs; one, the ground plan taken at the surface, and may be called the external plan or territorial map of the district; the other should be a plan or horizontal section made in the interior of the earth, and it is upon this that all remarkable geognostic objects are to be traced. That this section may be made in the most advantageous manner it should pass through the bottom of one of the draining levels, particularly through that which is traversed by the greatest number of galleries or subterranean shafts, and where consequently the strata are best known: for this reason it should pass through

 the

the deepest draining gallery or level of the district. This design should be entitled Internal Plan of the District, or rather Section or Plan of the Bottom of the Gallery. On these two kinds of plan should be marked not only the different kinds of rock and their extent, distinguished from one another by pale colours, but the different mineral repositories should be traced (beds, veins, &c.), attending to their direction, turnings, variations and size, as far as possible. On these plans should be marked the remarkable and known beds and strata of the rock, as the beds of porphyry, hornblende slate, quartz, limestone, and such like beds.

The vein formations must be distinguished by colours different from, and of a deeper shade than those that characterize the rocks, being particularly attentive to mark the intersections and derangements of the veins. On the external plan should be marked the situation of the vein. As the level on which the internal plan is constructed is driven upon the veins of the district, the plan of the level will contain that of the veins: that part of the vein through which the level runs might be represented by lines; and the rest of the direction, in case it should be already known by other works, may be given in dotted lines.

In

In the following way we may nearly represent the size of veins and other repositories. We can mark by an ordinary line such veins as are six inches thick, or under it; veins that are from six to 20 inches thick may be distinguished by a line the 16th part of an inch in thickness; from 20 inches to a fathom the line may be the 12th part of an inch thick; and if the vein be several fathoms thick, it may be represented by a line the eighth of an inch.

With respect to the colours to be given to the different mineral repositories and metalliferous veins, they will be regulated by the metal which is the principal object for working the mine. Thus for gold, yellow will be employed; for silver, red; for copper, green; for lead, blue. But as the primitive colours are not sufficiently numerous for distinguishing all the metals, we may employ the same colour which is made use of in one country to denote a different metal in another country: or to distinguish a metal we may make use of a shade of one of the primitive colours; for example, silver may be denoted by a yellowish red, and mercury by a bluish red.

To avoid any greater length, I omit the other details relative to the composition of the plans.

These plans, designed with correctness, will in their

their comparisons afford very great instruction, and mutually explain each other.

§ 104.

Such plans, for extensive districts, drawn upon one single leaf, would be greatly too large and inconvenient for general use; it therefore becomes necessary to subdivide them, and to design them on separate leaves, so that each may contain a fourth part; and in case of necessity, these can be laid together, and joined so as to form a whole. Each quarter may contain 500 fathoms in length, and as much in breadth. These different plans may be laid down in relation either to the direction of the principal veins, or to the true meridian of the place or the magnetic meridian, or to any point of the compass, provided the direction be determined. Each quarter may be divided into different compartments, to which separate denominations may be given, and different numbers attached.

This subdivision of a district into quarters will facilitate and greatly abridge the construction of general plans; for, after having once parcelled out the district into quarters, the plan of each can be taken up separately.

The plans of the quarters may also serve for the construction

construction of a particular plan of the ground belonging to any one mine.

Besides these two principal plans (the external and internal) of a district, we should also have for each principal vein, a plan or section made according to its direction and inclination, as well as a vertical one in profile. On the section may be marked all the workings which have been made into the vein, and a note should be kept of the successive progress of each, and of the ores which have been taken out.

§ 105.

The description of a mine district ought to contain an account of its external surface, of its situation, of its limits, of the rocks which it contains, as well as the relation they bear to one another: this description ought also to contain an exact account, both general and particular, or rather individual, of the different known and remarkable mineral repositories contained in the district.

§ 106.

In the general description of the mineral repositories, we must determine the age and distinctive character of each formation. When we give a particular

lar description of each of these repositories, it will be proper to place them one after another, in the same order that they were found in; they must be described in detail, and each particularly: we must take notice of every thing interesting, and which may have any influence on the future labours of working.

§ 107.

Each particular description of different mineral repositories, should be entitled *Account of the different mineral repositories of the district.* The title of the general description should be, *General and geognostic description of the district.* This last work will contain not only a description of the district in general, but also an account of the different mineral repositories which it contains. It will serve as a ground-work for all particular descriptions, and will present an outline of the district. It will be in a measure an abridgement of the particular descriptions, and this last will be the developement of it.

§ 108.

The order in which it will be best to arrange the particular descriptions is the following. The particular description of each repository ought to be contained

tained in a separate sheet properly margined and numbered; these sheets should be disposed in order in pasteboards. In large districts that are divided into quarters, the description and plan of each district, with a short note giving an account of the repositories which it contains, should be put together into one pasteboard.

§ 109.

As in this treatise, I am, properly speaking, only treating of veins, I believe I ought to confine myself to those particulars which demand attention in describing them; but what I say on the subject of veins will apply very nearly to other mineral repositories. The following things are to be observed in giving the description of a vein.

I. *The exterior relations of the vein*, which comprehend,

1. *Its position.* In this we have to remark

A. Its distance from certain fixed and known points.

B. Its direction.

C. Its inclination; that is to say,

a. Its angle of inclination.

b. The point of the compass towards which it inclines or dips.

2. *Its*

2. *Its magnitude.* Here we have to consider,

A. Its width.

B. Its extent, which includes

a. Its length.

b. The manner in which it terminates.

C. The determination of its course.

D. Its ramifications, if there be any.

II. *The internal state of the vein.* On this part we attend to

1. *The predominating ores and veinstones:* of these

A. We give a short oryctognostic description, and note

B. Their frequency.

C. The order in which veinstones and ores are found in relation to one another.

D. The different variations which occur in certain parts.

2. *The ores and veinstones* which occur in veins, though *in smaller quantity.*

A. Of these we must give a short oryctognostic account, and mark

B. Their frequency.

C. The order in which they are found.

D. Where.

E. Under what circumstances they occur.

3. What concerns the *metalliferous parts,* that is

to

to say, those parts of the vein where the ore occurs, as,

A. Their size.

B. Their richness.

C. Their frequency and distance from each other, and

D. The places where they are and have been found, noting whether they are increasing or diminishing.

4. *The other internal circumstances* of the vein, viz.

A. The druses.

B. The fragments of rock found mixed with the substance of the vein.

C. The borders (Besteg), and

D. Whether it adheres to the rock or not.

III. *The adjacent rock:* of which we remark,

1. The nature and quality.

2. The inclination of the strata.

3. The alteration on the roof or wall, which comprehends

A. The greater or less degree of decomposition it has undergone.

B. The metallic particles with which it is impregnated.

4. If it be rent, and in what manner.

5. The particular beds which occur in the rock. Of these

A. We must give a short oryctognostic and geognostic description.

B. The places where they occur.

C. The effect they have on the vein, and reciprocally.

IV. *The relation of the vein we are describing to the vein or veins which it meets, and reciprocally.* On each of these veins we must remark,

1. The point of meeting.

2. We give a short account of their direction, inclination, and depth; of the nature of the ores and veinstones.

3. We must describe the peculiarities presented by these veins in their intersections with the principal veins; that is to say,

If, after meeting, they continue together for some time.

If the veins that meet the principal vein traverse it.

If they are traversed by it.

If they produce ramifications, or

If they break it, or are broken by it.

If they derange it, or are deranged by it, and the magnitude of this derangement.

If they intercept and entirely stop the course of the principal vein, or if they are intercepted by it.

What influence and changes they produce on the nature of the veinstones and ores.

Such are the principal circumstances to be noticed in describing veins; it will also be of consequence to keep a note of the two following points.

V. Of the *principal operations* in the way of working, done in the vein which we are describing, as,

1. The excavations.

2. The works of trial which have been made on the vein, more particularly

3. The principal depth to which they have gone in the vein.

VI. With regard to the *size of the grant*; in which must be noticed,

1. The point where it begins.

2. The extent, as well as

A. The possessor, and

B. The length of it.

The particular description of a vein will be given in the account of the quarter where it *chiefly* occurs; in other places, where the same vein may again appear, it will be enough to refer to the description already given, only mentioning the principal changes which the vein may have suffered, as well as the works

works driven into it in the quarter which we are then describing.

§ 110.

A certain number of persons, to whom suitable instructions will be given, and whose labours will be suitably directed, must be charged with making a particular description of the veins found in each mine; this is the first thing to be done in order to obtain the materials which are necessary for completing the account of the mineral repositories, or rather the description of the particular repositories which ought to form the general account. Some one better informed should then be appointed to examine, rectify, and put in order these first materials; he must even go to the spot to examine more particularly the objects which appeared most interesting.

Such an account of mineral repositories requires much trouble, and a considerable time, to render it complete; but from the very commencement, every step made in the labour will be profitable and useful of itself. Likewise, as I have already said, it is only by adding from time to time the new observations arising from our labour, that we can hope to render it perfect. To attain this with more ease, it will be proper to order, that in every district, the officers of the

the mines give an account not only of every new operation in the working, but also of every change the vein suffers; so that in every thing done in the mine, whether in the way of working, traversing, or examining more particularly a repository already known, every observation may be shortly inserted in a journal kept for this particular purpose.

§ 111.

In drawing up an account of different mineral repositories, it will also be proper to make a collection of the minerals which are found in the district; this should be done geographically, that is to say, following the geographic order of the repositories; to each specimen should be attached a small note, in which is marked the repository and place where the mineral was found, the quantity of it, and the manner in which it occurs.

§ 112.

Such a collection, the plan and description of the district, form together a complete instructive whole. If our ancestors had left us such documents for two centuries past, or even for half a century, what advantage would it not have been of to us? From what doubts would it not relieve us? With what anxiety do do we not turn over the leaves of ancient chronicles in search of information, often very imperfect, obscure, and uncertain? With what pleasure do we not receive the least sketch or plan of some ancient mine? With what pains do we not rake up the old heaps of rubbish brought out of old excavations, to discover pieces which may afford us some idea of the substances which were formerly worked out? Yet between these documents, and those which we might obtain in the way pointed out in the preceding paragraphs, there is as much difference as between night and day. Would it not be an obligation, a duty, for us to collect and leave to future generations as much instruction and knowledge as possible on the labours carried on in our mines, whether it be in those that are still worked, or in those which have been given up.

CHAP. X.

AN ABRIDGED ACCOUNT OF THE PRINCIPAL METALLIC VEIN FORMATIONS WHICH OCCUR IN THE MINING DISTRICT OF FREYBERG.

§ 113.

I SHALL conclude this treatise with a short application of my theory to the mining district of Freyberg, by describing the different formations of metallic veins which I have observed in it. This description will not only serve to explain and elucidate my theory, but may also be regarded as the beginning of a description and more exact knowledge of the district of Freyberg. By it the practical miner of this country will be enabled to form a more correct knowledge of the nature of the repositories which he works. It will also assist in the practical study of veins those who come here to study mineralogy and

geognosy:

geognosy: those persons have the greater occasion for such an assistant in studying our veins, because they are much more complicated in their nature than the veins that occur elsewhere.

§ 114.

Under the name of *the mining district of Freyberg*, I comprehend that natural district in the middle of which the city of Freyberg is situated, and which contains several depots of metalliferous veins, if I may so express it, interwoven with one another. The mines of this district have for several centuries been an object of considerable importance, and they still continue to flourish. These mines are near Freyberg and the villages of Lossnitz, Hilbersdorf, Weissenborn, and Berthelsdorf; Brand, St Michel, Erbisdorff, and Langenau; Tuttendorf, Conradsdorf, Halsbrücke, Rothenfurth and Gros-Schirma.

This district appears to extend southward, near to Langenau, not quite so far as Gros-Hartmansdorff, but to Müdisdorf, and almost to Lichtenberg: on the east, as far as Weissenborn, and almost to Nauendorff, without going to Ober and Nieder Bobritsch: on the north it extends to Niederschöna, Krumhennersdorf, and stretches almost to Hohe-

Tanne,

Tanne, very near to Grosschirma: on the west it extends to Waltersdorf, Klein-Schirma, and Linde, so that the extent of this district is nearly two German geographical miles long, and more than one mile broad. But as it is very difficult to fix exactly the limits of a district, as well as of a depot of mines; and farther, as there are no determinate limits in nature, since formations, depots, and mine districts, gradually disappear; it may happen, that, beyond the limits which I have assigned, we may find some traces of the formations which belong to the district of Freyberg.

The jurisdiction of the council of mines at Freyberg, extends over several other small districts and depots of mines; such as the districts of Voigtsberg, of Gersdorff, of Memmendorff; the depots of Grund, Muntzig, Scharfenberg, Braunsdorff, Ober-Schöna, Fraüenstein, and several others. The depot of silver at Ober-Schöna, in the vicinity of Linde, belongs perhaps to the mining district of Freyberg, in the proper acceptation of the expression.

§ 115.

Within the limits I have just marked out, I have observed at least eight principal depots of metallic veins, without reckoning some others of less note.

These depots are perfectly distinct from one another, and, for the most part, contain several different metals. I shall describe the principal ores, following the order of their relative age, in so far as I have been able to ascertain it by my observations.

§ 116.

The first, and decidedly the most ancient of these depots, is a depot of argentiferous lead glance. On account of its richness, it is one of the most important in the whole district; since the earliest period of working the mines of Freyberg, it has uninterruptedly afforded a large quantity of lead and silver, and a smaller of copper; and it continues still to yield the same. It consists of

Coarse granular *lead glance*, containing from one and a half to two ounces and a half of silver in the quintal;

Common arsenic pyrites;

Black blende, in large grains;

Common iron and *liver pyrites*, and sometimes

A little *copper pyrites*, with

A little *sparry ironstone*.

The veinstones are chiefly

Quartz, sometimes

A

A little *brown spar*; and rarely

A little *calc-spar*, almost always crystallized.

Of all the fossils of this formation, *quartz* appears to be the oldest, and to have been first produced.

The *lead glance*, *black blende*, *iron*, *arsenic and copper pyrites*, appear, for the most part, to have been formed at the same time, but later than the *quartz*. If at any time there appear to be any difference in the time of the formation of these minerals, the *blende* seems to be the oldest, and the lead glance the newest. The *sparry ironstone*, and the *brown spar*, are, however, of still newer formation, for they are almost always found in the middle of the veins of this formation, and they frequently form druses. The *calc-spar*, which we but rarely meet with, and in small quantity, in the veins of this formation, is the newest fossil: its crystals cover the walls of the druses. Besides the minerals we have just mentioned, which are intimately mixed, and compose the body of the vein, there are sometimes found small portions of *copper pyrites* and *lead glance* of still newer formation. These two mineral substances, mixed with a little *quartz*, are almost always found crystallized on the *brown spar*; but they are always under the *calc-spar*, and consequently are older. The *copper pyrites* occurs in smaller quantity than the five other

other minerals which we have just spoken of; and it appears to me that they do not occur in all the veins of this formation. In some of them we meet with a little *grey copper ore*, which always occurs massive or disseminated, and mixed with the *copper pyrites;* sometimes it even approaches much to the nature of this last mentioned mineral. I am uncertain whether it belongs to this formation, or if it do not constitute a particular one.

The depot occurs most generally in the northern veins; it extends to the east and south of Freyberg, and consequently occupies the greater part of the district of Hohe-Birke. The size of these veins varies from six inches to two feet and a half; and they occur in great number. The principal ones are the western vein Abraham of Neue-Morgenstern; the eastern vein at Morgenstern; the northern veins Kirschbaum, Abraham, Thurmhoff, Joseph, Kuhschacht, Thomas, Kröner, Junge-Hohe-Birke, Jonas, and Junge-Mordgrube. Those in which *copper pyrytes* is chiefly found, are, the western vein Abraham, the eastern vein Morgenstern, the northern veins, Abraham, Kuhschacht, Hohe-Birke, Kröner, Junge-Hohe-Birke, and Jonas. Those that contain *grey copper ore*, are the northern veins, Hohe-Birke, Kröner, Junge-Hohe-Birke, Jung-Andreas, Jonas, and

some

some others in the mines of Junge-Thurmhoff, Rosinkrantz and Beschert-Gluck; but I have not yet been able to ascertain whether these last mentioned veins belong to the formation I have described, or if they do not constitute a particular formation.

There are two small depots of this formation beyond the Freyberg district, viz. in the Grund, and at Muntzig; the former occurs in a rock of *porphyry*, and the latter in *clay slate*. As this depot or formation occurs in these two varieties of rock, we may conclude that it is newer than them. It may perhaps be also found in different parts of the Erzgebirge; the depot of lead at Drehbach appears to belong to this. I am not in possession of a sufficient number of facts to be able to say any thing with certainty on the others. Nor can I say with confidence that this formation occurs out of Saxony; we know that *black blende* and *arsenic pyrites* occur in many parts of this country, more frequently than any where else.

§ 117.

The second depot of the mining district of Freyberg, which is a depot of *silver* and *lead*, is the most important depot of this district, in regard to the quantity of silver which it has yielded, and still continues

tinues to yield: the ores of which it is composed, are,

Lead glance, large and small granular, and very rich in silver;

Black blende, small granular;

Common *iron* and *liver pyrites*, and almost always

A little *arsenic pyrites*.

Further,

Dark *red silver ore*;

Brittle silver ore;

White silver glance, and

Plumose antimony ore.

The veinstones are,

Principally *quartz*;

Much *brown spar*, and often

Calc spar.

In this formation it is very easy to distinguish the age of the different kinds of minerals which it contains. The *quartz* is almost always the oldest, being constantly deposited on the walls of the vein. Its crystals, in hexagonal prisms, implanted at one of their ends, form the walls of the druses. Upon and in these walls, we find *black blende*, *arsenic pyrites*, *lead glance* and *iron pyrites*. The *blende* and *arsenic pyrites* appear to be the oldest formation. Next to them comes *brown spar*; after that the three ores of of silver, viz. *brittle silver ore*, *red silver ore*, *white silver ore*; to these succeeds *lead glance*, and sometimes a small portion of common *iron pyrites*. The lead glance appears to be the oldest of these five substances; and the three ores of silver seem to have been produced at the same time. Sometimes we find above these ores of silver, some very small and sharp pointed crystals of *quartz*. Lastly, We find the *calc-spar*, which when it does occur is the newest fossil, and always occupies the middle of the vein; and when there are druses, its crystals line the sides of them. The *plumose antimony ore* occurs but seldom in the veins of this formation, and then always in druses, which shows it to be the newest of the minerals of the formation. I believe it to have been contained in the same solution with the three silver ores, but to have been the last precipitated.

Although the minerals which compose this formation occur almost always together in the same veins, it sometimes happens that the *black blende*, the *arsenic pyrites*, the *lead glance*, and the *iron pyrites*, occur almost alone with the quartz and a little brown spar; at other times, on the contrary, the brittle silver glance, the red silver ore, the white silver ore, and the lead glance, occur with the quartz and brown

spar

spar alone; and when these last mentioned ores are in the same vein with the first, they almost always occupy the upper part of it. From this it appears that this formation might be subdivided into two others, of which the one that contains the silver ores is decidedly the newest. But as these two formations occur almost always in the same vein, I in the mean time consider them of one formation. If we wish to separate them, the first may be called the depot of lead glance rich in silver, and the second may be denominated the depot of red and white silver ore. The brown spar of this last formation has generally a flesh red, sometimes a rose red colour; when it happens to be crystallized, the crystals have the form of small common lenses. These two formations do not differ much in their age; they are older than the following, but newer than those that have been described in the preceding paragraph, which are the oldest in the district of Freyberg.

This second depot occurs in the south and south-west veins, and they are commonly narrow, from two to ten inches wide. They chiefly occur in the district of Brand; in the greater number of the veins of Himmelsfürst, Rosen, Donath, Gelobt Land, Altgrünzweig, Vergnute, Anweissung, Joel, Palmbaum, and Herzog August. Also in the mines

Neu

Neu Glück drey Eichen, Beschert Glück, and Jung-Himmlisch-Heer; at Alte Elizabeth, and, if I am not mistaken, at Krieg and Frieden. Beyond the district of Freyberg, but within the jurisdiction of the council of mines, this depot occurs again in the district of Voigtsberg, at the rich mine of Alte Hoffnung Gottes. I have not found it anywhere out of this country, but in Saxony.

§ 118.

The third is a depot of lead glance, containing but little silver; it has been worked from the earliest times, and contains

Lead glance, which yields nearly an ounce of silver in the quintal.

A great deal of *iron pyrites*, mostly the common kind.

A small quantity of *black blende*, and

Almost always a little *red iron ochre*.

The veinstones are

Quartz; sometimes also

Chlorite earth, mixed and surrounded with clay.

The *lead glance* of this depot is often in lamellar concretions.

This depot appears to be much newer than the preceding. It is only found in the northern veins,

 and

and in such as are of moderate size. It is chiefly found in the district of the city, and in that of Halsbrükke. To this formation belong the veins in the mine of Priesterlicher Glückwunsch; and probably also the veins of Anna Fortuna, the northern veins Dreyfaltigkeit, and Nachtigal near Tuttendorff, the veins of Himmelfarth Christi, the northern vein Rothgrube, and several others. All the veins of this district, called by the ancients veins of pyrites, belong to this third depot, according to every appearance.

§ 119.

The fourth depot of metalliferous veins in this district is also a depot of lead glance containing but little silver: It is much newer than the preceding. The minerals which it contains, are,

Lead glance, almost always containing from a quarter to three quarters of an ounce of silver in the quintal.

Radiated pyrites; sometimes

Brown blende in small quantity.

The veinstones are very distinct, and consist of

Heavy spar, almost all the varieties.

Fluor spar, yellow, pale blue and white.

A little *quartz*, and rarely

Calc spar.

This

This depot occurs most commonly in veins that are from a foot to two fathoms wide: they have generally a western direction. It is chiefly found in the district of Halsbrücke; it forms a vein of considerable magnitude, called *Halsbrückner Spath*, as well as its two branches, one of which runs westward to Kurprinz, and the other east as far as Lorenz Gegentrum. This same depot also occurs in the veins *Isaak Freudenstein*, and *Kom Siegmit Freuden*, and in several other veins, which, branching off at this place, and taking a southern direction, stretch under the town. Amongst these are the western veins at Morgenstern, a branch of the western vein *Abraham of Neue Morgenstern.* This formation occurs also in other districts, which hold of the council of mines at Freyberg; among others, we meet with it in the districts of Gersdorff and Memmensdorff. It occurs in our high mountains near Tschopau, at the mine Dreyfaltigkeit. In all these three places it occurs in western veins, which at Gersdorff and Tschopau are very large. I suppose it also occurs at Annaberg, in the mine Galiläische Wirthschaft, and in several others. It is also met with out of Saxony, as in the greater number of the veins at the Peak of Derbyshire in England, and at Gisloff in Schonen, a province of Sweden. In the former of

these

these countries it is contained in transition limestone rocks, a proof of its newness.

§ 120.

Along with the formation we have just described, there occurs another which consists of

Grey copper ore of a light grey colour, and rich in silver.

A little *copper pyrites*, and

Lead glance, small and fine grained.

This formation occurs in

Hornstone passing to quartz;

A little common *heavy spar*, and

A little *fluor spar*.

It seldom occurs but in the veins of the formation just described, and there in particular ramifications: it is nearly of the same age, but appears to me to be a very little older.

It is chiefly met with in the large vein *Halsbrückner Spath* and its continuations; in the veins near *Gersdorff*, and near *Tschopau* in the *Dreyfältigkeit*.

§ 121.

The fifth depot of metalliferous veins in this district

strict is of native silver, silver glance, and glance cobalt: it contains

Native silver, capillary, dentiform, and in membranes;

Silver glance;

Glance cobalt, commonly the reticulated; sometimes also

A small portion of *grey copper ore*;

Lead glance, rich in silver;

A little *brown blende*, fine grained, and

Sparry ironstone, fine grained.

The veinstones are

Heavy spar in a state of disintegration;

Fluor spar of a violet blue colour, and fine grained.

In the district of Freyberg, this formation always occurs in the intersections of the southern and western veins; (the former contain most commonly the first, and the others the fourth formation which we have described), sometimes it is found even in the middle of the western veins, and the working of it has in some places been very profitable. The veins of this formation are chiefly to the east and north of the city, namely, in the mines of *Himmelfahrt*, *Segen Gottes*, *Schlöschen*, *Morgenstern*, and *Neue Morgenstern*. The considerable profit formerly derived from

from Morgenstern, arose from this depot: the ore occurred at the intersection of the western veins *Gutmorgen* and *Silber-präsent*, with the eastern vein of *Morgenstern*. This formation still furnishes the considerable revenue arising from the mines of Himmelsfahrt and Neue Morgenstern; even in former times it must have yielded a large quantity of silver, and it probably was the one which, according to Albert the Great, contained the capillary silver which was worked in the middle of the 13th century *. I have observed this formation in the *Ober-Erzgebirge* in the district of *Marienberg*, among other places at *Dreyfaltigkeit* near *Tschopau*, at *Glücksgarten* in *Hopfgarten* below *Wolkenstein*; in the district of *Annaberg* at *Markus-Röhling*, *Galiläische Wirthschaft*, as well as in other mines near *Schletau* and *Scheibenberg*. I am uncertain if it occurs anywhere but in Saxony.

The grey copper ore, rich in silver, together with

* De Mineralibus et Rebus Metallicis libri quinque, auctore Alberto Magno. 12mo. Colon. 1669, p. 280. 281. in the following words. "Invenitur enim (sc. argentum "nativum), in loco Theutoniæ qui dicitur Wrieberg quod "sonat liber mons, aliquando molle sicut pultes tenaces, "et est purissimum et optimum genus argenti."

with the lead glance and brown blende which I have met with in this formation, may, as I strongly suspect they do, constitute a distinct and particular formation.

§ 122.

The sixth depot of metalliferous veins is a depot of native arsenic and red silver ore.

It contains

Chiefly *native arsenic* and light red silver ore;

Sometimes a little *orpiment*;

Rarely a little *copper nickel*;

Glance cobalt;

Native silver;

Lead glance;

Iron pyrites, and

Sparry ironstone.

The ores occur in

Common or lamellar *heavy spar*;

Green *fluor spar*;

Calc spar, and

A little *brown spar.*

This formation occurs at Freyberg, in the intersections, or in the middle of veins, particularly those of the two preceding formations. They are found in the middle of the vein in the mine *Kurprinz.* I have seen

some

some traces of it at *Isaak.* It also exists at *Bescher-Glück*, but I do not know in what manner it occurs. I have an indistinct recollection of having also seen it in the mine *Himmelfahrt* and *Gott-mit-uns.* In the districts within the jurisdiction of Freyberg, I only know of it being found at *Gersdorf;* but of this I have only seen one small piece. It is chiefly found in *Ober-Erzgebirge*, and principally at *Palmbaum* near Marienberg, where it occurs along with a little lead glance and molybdena. It is also found at *Bierschnabelstotlen* near Annaberg, and at *Herzog Karl* near *Ehrenfriedersdorf.* The numerous ores of silver, which in former times were worked out at *Ehrenfriedersdorf* in the *Sauberg*, as well as in the neighbourhood at *Klingelschlägel*, seem, according to the specimens I have seen, to belong to this sixth formation. It still is found in the vicinity of *Barenstein* and of *Weisenthal*, chiefly in the mine *Kinder Israel.* It was formerly found at *Neu-Leipsiger-Glück* near *Johanngeorgenstadt.* Out of Saxony, it occurs chiefly at *Joachimstahl* on the *Huber.* I believe it is also found in the mines of *Sante Marie* in *Alsace.* I am uncertain if I ought to consider the glance cobalt (with the copper nickel and native silver), which I have ascribed to this formation, as really belonging

to

to it, or as forming a part of the preceding formation; I incline towards this latter opinion.

This sixth formation, as well as the fifth, is undoubtedly of later formation than the fourth, for both of them are only met with in the intersections, or in the middle of the veins of the fourth formation; but they always occur separately. I cannot yet affirm, with certainty, which is the newest of the two; to me, however, it appears to be the last, viz. that consisting of native arsenic and red silver ore. For I believe these two formations sometimes occur together in the mines of *Ober-Erzgebirge*, and in them I have found red silver ore resting upon the glance cobalt.

§ 123.

The seventh depot of the metalliferous veins of the district of Freyberg, is a depot of red ironstone, and contains only

Ochry red ironstone, or red hematite;

A little *iron glance;*

Quartz, and

A little *heavy spar.*

This formation, which is but of little consequence in working the mines of Freyberg, occurs in the vicinity of *Losniz* and of *Waltersdorf;* in the mines

Anna Fortuna, *Vergnügler Bergman*, and *Johannes* of *Lerchenberg*: farther, in the valley of *Münzbach*, and in other mines to the eastward and south of the town. It commonly occurs in the upper parts of the vein. It is also found in the mines of *Zwölf Schlüssel*, *Roth und Weisser Löwe*; and I believe it was met with long ago in those of *Junge-hohe-Birke*, *Junge Andeas*, *Kröner*, and very likely at *Junge Thurmhof*. This is certainly one of the newest formations, as appears from the places which it occupies in the veins. This formation probably affords the numerous and considerable veins of red ironstone, which, commencing in the *Obergebirge* at *Gieshübel*, stretch across *Voigtland*, and extend as far as *Wolkenstein* and *Ehrenfriedersdorf*. It is not improbable that the red colour of the gneiss, in some parts of the surface of the mountains of Freyberg, is produced by this formation of red ironstone.

§ 124.

The eighth depot of the metalliferous veins of the district, is a depot of copper, consisting of

Copper pyrites;

Mountain green;

Malachite and *red and brown iron ochre*, with

A

A little *quartz*, and

A little *fluor spar*.

This formation, which is of little importance, occurs about a league on the east side of the city, chiefly at *Conradsdorf* in the *Johann-Georgenstadt*, *Neubeschert-Glück*, *Lorenz Gegentrum*. I cannot say with certainty if the copper veins at *Johannes* on the *Höglez-Höhe*, not far from *Altenberg*, as well as those in the Stockwerke of *Leisson*, and in the mine of Fortuna near Grunthat, which contain the same ores, belong to this formation or not.

§ 125.

Such are the principal metalliferous veins in the mining district of Freyberg. If we consider the number and richness of the veins contained in this district, we shall not be astonished at the immense treasure which our veins have already yielded; and we may indulge the well-founded hope of successful operations in time to come.

Finally, I have no doubt of being able to add to these formations of metalliferous veins, some others less remarkable, and not so rich. For example, I think I have seen in the mine Himmelsfürst, two formations of silver ore, which do not occur in the other mountains.

One

One of these formations contains

Native dendritiform silver, in

Common *heavy spar*, and

Brown spar.

This formation occurs also at *Wit-tichen*, in the Black Forest, in *Swabia*, in the country of *Fürstenberg*, and at *Sante Marie* in the mines of *Alsace*.

The other formation contains

Native dentiform silver ;

Silver-glance ;

A little *brown blende*, and

Sparry ironstone.

This occurs in and with

Calc spar, and

Brown spar.

This formation, I believe, is to be found at *Ratiborschiz* in *Bohemia*, and *Priester* at *Schneeberg*.

I do not yet know to what formation the following belongs, viz.

Light *red silver ore*, partly crystallized, and partly in the form of membranes, which was formerly, and still is found at *Beschert Glück*, *Junge Thurmhof* and *Kröner* ; I am induced to think it belongs to the fourth formation, namely that which contains *native arsenic* and pale *red silver ore*. It is most commonly found in veins that are of a ferruginous nature.

Further,

Further, we sometimes find in several of the veins of the Freyberg district, chiefly in the mines *Johannes* on the *Lerchenberg*, and *Lorenz Gegentrum* and *Jonas*,

Variegated copper ore, with

Copper pyrites, and even

A little *copper glance*.

But I do not know whether this formation should be considered to belong to the preceding ones, or if it is a distinct formation.

I imagine, however, that beside the three depots of lead which I have mentioned, there occurs a fourth, which seems to be newer than any of the preceding, and consists of

Lead glance, in lamellar concretions, rich in silver;

Compact lead glance;

A little *iron pyrites*;

Black blende, and

Sparry ironstone.

It always occurs in loam.

I believe it occurs at *Holewein*, in the vein *Heinrich*, at Morgenstern in the vein *Harnisch*, and at *Himmelfahrt* in the vein *Fuschglück*; all these veins which are of trifling size, have a southerly direction.

They

They still speak of tinstone, which is said to have been formerly worked in the south-east part of the district of Freyberg; but I have not yet been able to observe the smallest trace of it.

I have already remarked, that the second, fourth, and fifth depots contain some ores which appear to be of particular formations.

I take no notice of veins which only contain stony substances, such as quartz, brown spar, heavy spar, and clay; because I only wish to treat of formations of metalliferous veins. Besides, I am not in possession of a sufficient number of data to be able to say any thing certain with respect to them.

§ 126.

Such is the summary of the principal observations which I have made on the district of Freyberg; the knowledge of which I have acquired from repeated observations made myself during 30 years. I have obtained it by frequently descending into the interior of mines; by a strict examination of the ores which I have seen in different collections, on the authenticity of which I could rely; and finally, in getting every imaginable information. I flatter myself that the exposition which I have given of the state of the mining district of Freyberg, will not only serve as a model for similar investigations, but that it will excite the patriotic efforts of the Saxon miners, and induce them to augment and complete the geognostic and mineralogical tables, of which I have only given a sketch.

§ 127.

I cannot refrain from saying, that in the country adjoining to the Freyberg mining district we find other three particular and very distinct metalliferous depots.

The first is a depot of red silver ore that contains

Deep red silver ore, which occurs in

Quartz passing into *hornstone*, which sometimes has a green colour near the walls.

This depot occurs in the mines *Daniel*, *Alte Hoffnung*, *Christ bescherung* near *Voigtsberg*, at *Braunsdorf*, and in several other places.

The second is an inconsiderable depot of antimony, and consists chiefly of

Grey antimony ore in

Quartz.

This occurs in the mines *Alte Hoffnung*, at *Klein-Voigtsberg*, *Neue Hoffnung*, *Siegfeld* and *Vertragliche Geselschaft* at *Braunsdorf*: it is found along with the former, but seems to be rather newer than it.

The

The third is the depot of lead and silver at *Scharfenberg*, and contains

Lead glance, fine granular, and rich in silver;

Yellow and red blende;

A small quantity of common *iron pyrites;* sometimes

A little *native silver*, and

Silver glance,

All of which occur in

Brown spar, coarse and fine granular;

Quartz; and even

A little *calc spar.*

This formation does not occur in any other part of Saxony; but it is seen at *Kapnik* in *Transylvania.* Perhaps it may contain a small proportion of gold.

SUPPLEMENT.

SUPPLEMENT.

§ 128.

SINCE the impression of my treatise, I have had an opportunity of making some observations which I shall here add, because they will serve to explain my new theory of veins, and even afford a proof of my doctrine.

In Crell's Chemical Annals, I have described a repository of wacke, known by the name of *Buzenwakke*, which occurs in the gallery *Barbara* at *Joachimstahl*. This wacke belongs to the trap formation, but it occurs in a primitive rock of mica slate and clay slate; the repository descends to a depth of 150 fathoms; it contains semi-petrified trees, the bark, branches, and leaves of which are still entire. This repository bears a strong resemblance to veins: it undoubtedly was an enormous rent formed in this high primitive mountain, which has afterwards been filled up from above with wacke. What revolution could produce this rent? And what one could have afterwards filled it up?

§ 129.

M. de Gruner of Berne, one of my most distinguished pupils, and a particularly good geognost, communicated to me a few years since a very interesting observation, viz. that in the Alps of Switzerland, namely, in the *Valais*, there are narrow valleys which cut and separate the ridges of the mountains, which appear to be nothing but great rents. A short time after I found the same observation contained in the mineralogical letters of Ferber.

Something of the same kind occurs in Derbyshire: Whitehurst, in his Theory of the Formation of the Earth, says, that such a rent is to be seen near Matlock, the bottom of which is filled with the debris of the rocks which constitute the mountain, and the upper part is the valley through which the Derwent flows.

§ 130.

M. de Trebra, the present captain-general of the

Saxon

Saxon mines, in the Gottingen Magazine, gives a short and instructive description of a cavity or druse discovered in 1785 at *Andreasberg*, in the vein *Fünf-Bücher-Mosis*. It is two fathoms and a half long, and in some parts 30 inches wide. It was not only lined with the most beautiful crystals of calc spar, but it also contained large pieces of this fossil, either reposing on one another, or projecting into the interior of the cavity: these pieces were covered in a similar manner with crystals which in part connected them with each other.

§ 131.

I was lately favoured by one of my friends with an extract made from Fichtel's Mineralogical Observations on the Carpathian mountains: this extract contains three examples of petrifactions found in the metalliferous veins of Hungary and Transylvania. I shall give it just as it was sent me *.

§ 132.

---

* Mineralogische Bemerkungen von den Karpathen, von Joh. Ehrenr. von Fichtel. 8vo. Vien. 1791, Part. I. pages 48. and 49. The whole passage deserves to be quoted. It is the following.

" I saw at Kremnitz a fossil brought from the mine
" *Stadthandlung*; it was a foliated fungus of the size of a
" nut,

§ 132.

I would recommend it to every mineralogist who is desirous of becoming acquainted with the nature of veins, to study those chiefly which occur in the mountains

---

" nut, the parallel folia of which contained a round ball " enclosed within them. This fungites and the ball were " composed of dark brown sparry ironstone (not brown " spar, for it contains no manganese) ; it occurred in cry- " stallized quartz which was found in decomposed grey- " stone ; besides the fungites, the whole surface of the " fossil is covered with larger and smaller rhombs of sparry " ironstone of a beautiful gold yellow colour.

" At Schemnitz, I also saw, but did not learn from " what mine it came, a bivalve shell about the size of a " hazel nut resting upon quartz, and this again on a de- " composed greystone; the two valves were separated from " one another in a perpendicular direction, perfectly en- " tire, and with a sharp point very much bent. The shell " is bent, very concave ; its substance thick, smooth, much " weathered, and appears to belong to the cordiform " shells.

" If to these two petrifactions be added the madrepore " mentioned by Aulic Counsellor Born, and also a snail in " my possession found in a gold vein in Transylvania, we " have four incontestable examples of petrifactions occur- " ring in the metalliferous veins of Hungary and Transyl- " vania."

In

mountains called Erzgebirge, in Saxony and Bohemia, (principally at Joachimsthal); and also those that occur in the Hartz, and in Derbyshire in England. Perhaps the mountains of Cornwall, and of Kongsberg in Norway, may also furnish some interesting observations.

In the concluding part of this passage, the same author makes the following remark.

" No one acquainted with the nature of rocks will consider this mountain as of Neptunian origin; its substance does not contain the least vestige of marine bodies, although they do occur in the veins which run through it."

FINIS.

APPENDIX

BY

THE TRANSLATOR.

APPENDIX.

NOTE A. § 3. p. 3.

ALTHOUGH Werner says, in this §, that veins always traverse or cut the strata of the rock in which they occur, it deserves to be noticed that the branches of veins sometimes run parallel with the strata which are traversed by the principal vein; and that in some instances the principal vein itself, in part of its course, runs parallel with the strata. Such portions and branches of a vein may be confounded with true beds, which shows the necessity of attending carefully to the orictognostic distinction of beds and veins as they occur on the great scale. We will in general find that rocks or minerals in beds are differently arranged, and have different orictognostic and geognostic relations from the same rocks and minerals in veins; and that in these two species of repository they will be accompanied with different minerals.

NOTE B. § 25. p. 46.

In addition to the theories enumerated and examined in the text, two others have appeared since the publication of this treatise.

1. A particular theory of veins sketched by Dr Hutton, in his work, entitled "Theory of the Earth", and lately illustrated with great ability by Professor Playfair, in his work on the Huttonian Theory: this theory is fully examined by Professor Jameson in his Elements of Geognosy.

2. A New Theory of Veins by Patrin, well known by his mineralogical observations on Siberia. It is fully detailed under the article *Filon*, of the Nouveau Dictionaire d'Histoire Naturelle.

Besides the different works treating of veins mentioned in this treatise, the following may be consulted with advantage.

1. Charpentier Beobachtungen über die Lagerstätte der Erze Hauptsächlich, aus den Sachsischen Gebirgen. Ein beytrag zur Geognosie. Met Kupfern, 4to.

This interesting work is divided into four books. The first book treats of the strata and beds of mountains, in so far as they are to be considered as repositories of ores. The second book treats of those mineral repositories denominated *veins*. The third book treats of the passage of veins through different species of rocks, as also of the change they

they induce on the neighbouring rocks. The fourth treats of veins filled with stony substances of different kinds.

2. Beiträge zur Naturgeschichte der Gänge, von Johann Karl Friesleben, in von Molls Jahrbücher der Berg und Hutten Kunde for 1800.

3. Bemerkungen über der Harz von Johann Karl Friesleben, 2 vols. 8vo. 1795.

4. Sammlung Mineralogischer und Bergmanischer Abhandlungen, von F. Mohs. Wien. 1804.

5. Reuss Mineralogische Geographie, von Böhmen. Dresden 1793 to 1797, 2 vols.

6. Von Buch Geognostische Beobachtungen. Berlin 1802, 8vo.

7. Jameson's Mineralogy of the Scottish isles.

8. Williams's Natural History of the Mineral Kingdom, 2 vols. 8vo.

This work is divided into three parts. The first part treats of the natural history of the strata of coal, and of the concomitant strata. The second part treats of the natural history of mineral veins and other beds and repositories of the precious and useful metals. The third part treats of the natural history of the prevailing strata, and of the principal and most interesting phenomena upon and within the surface of the earth. Several years ago the first part of this work was translated into German, accompanied with many valuable illustrations and additions; more lately, as we are informed, nearly the whole work has been trans-

lated

lated into the German at the requeſt of Werner, and is accompanied with many valuable notes by the translator Hennrich Meuder, a pupil of Werner's.

Note C. § 31. p. 51.

Although the assistance which the crossing of veins affords us in determining the relative age of each has been much overlooked by mineralogists in general, it had not escaped the observation of Pryce, who notices the circumstance in his work giving an account of the country and mines of Cornwall. Theſe observations must have been unknown to Werner; for it cannot be supposed that an author who has showed so much anxiety to confer on every writer the merit to which his remarks entitle him, would have passed over this circumstance had the work fallen under his perusal.

As the passage is one of much importance, from confirming this part of the Wernerian theory, I shall give it in his own words.

" Because the Cross-Gossans, or Cross-Flookans, run
" through all veins of opposite directions, without the
" least interruption from them, but, on the contrary, do
" apparently disjoint, and dislocate all of them ; it seems
" reasonable to conclude, that the east and west veins
" were antecedent to cross veins ; and that some great
" event, long after the creation, occasioned these trans-
" verse clefts and openings. But how, or when, this
" should

" should come to pass, we cannot presume to form an " adequate idea."

Vide Pryce's Mineralogia Cornubiensis, p. 101.

Note D. § 40. p. 61.

In many of the sandstone districts in Scotland, particularly in the Orkney and Shetland isles, there are numerous and very extensive deep and wide rents and fissures still open.

Note E. § 48. p. 74.

The occurrence of coal in veins is a rare phenomenon. I am happy to be able to give an instance of it observed by a miner of great sagacity and skill, the late Mr Williams. In his valuable, but little known work, entitled Natural History of the Mineral Kingdom, there is the following description of veins of coal at Castle Leod.

" I have seen coal in the cavities of mineral veins at " Castle Leod, in the highlands of Scotland. This extra- " ordinary coal is lodged in common rake veins, or per- " pendicular mineral fissures in the mountain rocks, far " from any strata of coal, or of the coal metals. The " veins at Castle Leod open into bellies or concavities of " different lengths and capacity, and close again, or check, " by the sides of the veins coming together; and they " have every character and description common to good " rake veins, and the coal is lodged in the cavities of

" them

" them exactly in the same manner as lead, copper, &c. " are found in such veins in other places. I got some coal " dug out of one of these veins, which I burnt in the " house: the coal was very soft and tender, being very " near the surface; but it was perfectly clean and bright— " about one foot thick where I opened the ground, but it " grew thinner towards the ends of the belly or concavity " of the vein, and it soon dwindled away to nothing, and " twitched out entirely; and I saw it open again in several " places in the bearing of the vein. This coal was exceedingly fat, so as to run together strongly in the fire " like the best of the Newcastle coal. I look upon this " phenomenon in the natural history of the mineral kingdom as a great curiosity. I call it a phenomenon, because it is an extraordinary appearance quite out of common experience. I did not know any thing about coal " in such a situation and circumstances until I saw this. " I had never heard nor imagined any such thing until " this offered itself to my consideration. I was at all due " pains to investigate this phenomenon thoroughly, so as " not to leave out any circumstance which should be " considered, and so as not to be mistaken in any " point.

" There are several mineral veins at Castle Leod running parallel to one another upon the north-east side of " a pretty high and rocky mountain; and there are some " lesser strings branching out from the principal veins.

" There

" There is coal found in three or four of these veins, " part of which had been wrought out of several bellies of " those veins, as far as they could go down for water, before " I was there; and as I had no apparatus for drawing the " water out of the old works, I was obliged to open new " ground farther forward upon the bearing of one of the " veins, out of which coal had been wrought.—When I " opened the ground the coal was only about one foot " thick, and it only continued for a few yards in length at " that thickness, before it began to dwindle away by de- " grees, and it was soon squeezed out entirely at both " ends of this little belly; however, I saw evidently, that " some of the bellies or concavities of the veins, out of " which they had formerly dug some coal, were wider " than the one which I opened. Some of them were up " to three or four feet wide between the sides of the veins; " and the concavities, out of which they had dug coal, were " also somewhat longer than the one I opened. But they " had chosen the best places for their trials, which was " not difficult, as the coal was to be seen in the veins " at the surface of the ground. I saw coal in several " places in the checks or twitches of these veins between " the open bellies, not above one inch in thickness.

" The circumstance of these coal veins opening into " bellies, or concavities, between the sides, and twitching " again by the two sides coming close together, is common " to all mineral veins of this description, though the con-

" cavities

" cavities of the veins at Castle Leod are remarkably " short; and the twitches, or close parts, are also short; " these being a species of piped rake veins, the charac- " ter and description of which is, that they open and close " again at short distances in the line of bearing, and the " pipes or concavities put down in the dip of the vein pa- " rallel to one another."

Note F. § 49. p. 75.

Granite, sienite, and greenstone, are rocks which bear considerable resemblance to each other; it is therefore of some consequence to attend to the nature of each, that we may be able to distinguish them, not only because an inattention to this circumstance has frequently led into error, but because without attending to it, it will not be possible to give an exact account of any tract of country where they occur. The author has, in a note to § 49. pointed out in a very perspicuous manner, the difference between sienite and greenstone; to these granite also bears a strong resemblance; they differ, however, very essentially, not only in their component parts, but in their geognostic situation and relations.

Granite, the oldest of the rocks which compose the crust of this earth, either forms the lowest bed of a mountain, or constitutes the whole mass of it, around which the other rocks are wrapped in the mantle-form. It is an aggregate rock composed of felspar, quartz, and mica, in different

ferent proportions, and these vary much in the size of their grains. The felspar is generally most abundant, and the mica in smallest proportion. It is often stratified. We never find the oldest granite constituting the mass of a vein, although the second and third granite formations frequently do. The third or newest granite formation sometimes is met with in overlying formation.

Sienite is an aggregate rock composed of felspar and hornblende; sometimes it contains a small admixture of quartz or mica: these however are only accidental, and when they are present it is in small quantity, or in particular parts, without affecting the general nature of the rock. Hornblende is the characteristic ingredient which serves to distinguish it from granite. It occurs always in unconformable and overlying stratification, or in veins. Its beds always rest upon the old or primitive rocks: we never find it under any of them. When it constitutes veins, they always traverse the beds of the primitive rocks. It is never traversed by veins of the primitive rocks, except by those of the third granite formation.

Greenstone is a compound aggregate rock consisting of hornblende and felspar, in small grains, which are frequently so minute that the rock is almost destitute of the granular structure, and the constituent parts cannot be distinguished. The hornblende always predominates, and gives to the rock the green colour from which it derives its name. It sometimes occurs stratified. This rock is found in all the different classes; sparingly in the primitive, more abundantly in the transition, largely in the floetz, and very copiously in the coal formation; in this last, veins of it are frequent, forming what have in this country been called whin dykes, a name destitute of any particular application, and which ought therefore to be discontinued.

From this variety and universality in the occurrence of greenstone, mineralogists who did not consider it with attention have frequently mistaken its nature; and it has not unfrequently been confounded with, and described as old granite. But the presence of hornblende, which never, even in the most minute proportion, is found to enter into the composition of old granite, establishes a marked difference. Mica, it is true, sometimes occurs in greenstone, more particularly in the transition kind; and may have occasioned these errors, errors which, however, we trust, will disappear with the advancement of the geognosy. When mica does occur in greenstone, it is to be considered as an adventitious occurrence, not affecting the true nature of the rock, which will be easily ascertained by a more particular and extensive examination of the bed.

We have, in this country, an interesting variety of transition greenstone in beds, containing an admixture of mica, in pretty large quantity, which has made them be considered by some as granite. The beds to which I allude occur in Lammermuir hills, near Priestlaw, of which there are

some

sections made by Fassnet burn. But an examination of the nature of these beds, and their alternation with grey wacke, convinced me that they are of transition greenstone. Since then this tract of country has been most minutely examined by Dr Ogilby, who presented to the Wernerian Natural History Society an excellent memoir on the subject, in which he pointed out and illustrated by specimens the true nature of these beds and the different veins which traverse them.

Besides the rocks noticed, there is a variety of greenstone which differs somewhat from any of them. This has been denominated by Professor Jameson *sienitic greenstone*. In it the felspar preponderates, and is commonly of a flesh red colour.

A consideration of the particulars mentioned in this note, shows the utility of the study of orictognosy as a preparatory step to that of geognosy; for without a competent knowledge of the nature and characters of the different simple fossils, the geognost will not be able to communicate to others what he may observe in the different mountain rocks, which to the speculative naturalist, form an object of so much curiosity, and to whose accurate and faithful detail so much must be trusted in the investigations in which he is concerned.

NOTE

Note G. § 49. p. 75.

Besides the rocks enumerated in the text, as occurring in veins, the following have also been observed by Professor Jameson, viz. *porphyry slate*, *pitchstone*, and *sandstone*.

Note H. § 52. p. 83.

Bedded veins are those into whose structure a variety of different fossils enter. They are what have by some been denominated *stratified veins*, but as the term stratification applies only to rocks whose substance is uniform, but separated into different layers or strata by parallel rents, it cannot with propriety be applied to veins that are composed of beds of different fossil matters.

In bedded veins the beds are arranged in a determinate order which corresponds on each side, and are parallel to the walls of the vein. The part which lies next the walls has been first formed, consequently is the oldest; the others have been formed in succession towards the centre, which is the last formed, and therefore the newest of all. In consequence of this deposition of the successive beds from the walls to the centre of veins, the last formed beds contain the impressions of the crystals of those that were first formed. The centre of bedded veins often remains unfilled up, and are then drusy cavities, which are

frequently

frequently lined with crystals of the last formed bed, which, in the empty space, and tranquil state of the solution in the vein, have shot into regular forms. The greater number of the beautiful specimens of crystallized minerals which we meet with in the cabinets of the curious, have been formed in this way in the drusy cavities of veins.

We meet with many instances of veins of this kind. In the Freyberg district they are, as is mentioned in the text, numerous. In this country they are not unfrequent. In the island of Arran we find many examples of them: there they are composed of greenstone, porphyry slate, pitchstone, wacke, &c. arranged in different, but always in regular and determinate order.

NOTE I. § 58. p. 89.

What Werner here alludes to by the intimate "connection which takes place between veins and the rock," applies not so much to true veins as to cotemporaneous veins which are but little noticed by him in this treatise. It was not till after the publication of it that his attention was more particularly directed to them. I have therefore judged it proper in this note to point out the most material points in which *cotemporaneous* differ from *true* veins, referring for a more detailed account of them to a paper on this subject by Professor Jameson, which I hope will soon appear in the Transactions of the Wernerian Natural History Society.

Cotemporaneous

Cotemporaneous veins have most generally a tortuous form, and frequently branch off in a variety of directions; they are always confined to one bed, and are never observed to traverse the different beds of a mountain; they are generally enclosed, and are not separated from the mass of the rock in which they occur by walls, like true veins, but are intimately connected with it, indeed they mutually pass into each other. Their nature approaches very near to that of the rock in which they occur. In the compound or aggregate rocks they commonly consist of one or more of the simple fossils of which these rocks are composed, sometimes of all the ingredients, either in proportions differing from the rock, or varying in the size of the grains, which, to a casual observer, may be productive of error, and lead into mistakes which are apparently at variance with the geognosy. They occur in all the mountain rocks, though they are most numerous in those of the compound kind. They are particularly frequent in some of the old primitive rocks; *granite* often contains many cotemporaneous veins of quartz, and felspar either singly or mixed together in different proportions; in *gneiss* they occur in great abundance, and when their structure approaches more to the granular than the slaty, which is frequently the case, they bear a great resemblance to granite, and have been by some considered as granite; but we know that veins of old granite never occur in any rock of more recent formation, and that when we do find veins of granite in gneiss

gneiss or mica slate: they are of the *third* granite formation, which only presents itself in overlying formation, or in veins. Such cotemporaneous veins are frequent in many places. I may mention them as occurring in the island of Arran, at the Tor-nid-neon, near Loch Rausa, in Glen Sannicks, and many other places of that island, which affords so many interesting particulars to the geognost. In greenstone of every formation, we frequently meet with cotemporaneous veins of felspar, hornblende, and sienitic greenstone, which often run a long way in the beds of the common greenstone, instances of which are to be seen at Salisbury Craigs in the vicinity of Edinburgh.

Besides, these cotemporaneous veins which, as mentioned above, are composed of materials which do not differ from the nature of the substances composing the rock in which they occur, we often meet with them consisting of materials differing a good deal from the rock. Of this kind are veins of steatite which are so often seen in serpentine, and of calc spar in limestone. And in the floetz rocks we very commonly find cotemporaneous veins of quartz; instances of which occur so frequently in most of the mountain rocks that they require no specification.

NOTE J. § 60. p. 90.

In addition to what has been said in regard to the thickness or width of certain veins in this §, we may add that veins of greenstone sixty feet wide occur in Arran; and

Professor

Professor Jameson mentions in his Geognosy veins of porphyry slate 140 feet wide that occur in the same island. It is worthy of remark that all these very wide veins in Arran occur in strata of sandstone which are nearly horizontal. It would seem that the widest veins generally occur in the most horizontal strata.

NOTE K. § 69. p. 119.

It is now no longer doubted that the same fluid may contain in solution different saline substances at one and the same time; and without any great stretch of analogy, the same thing may be extended to earthy substances. This idea I find to be entertained by a naturalist whose authority must give weight to any opinion; I allude to the Count de Bournon, who expresses himself as follows. "When a fluid holds in solution the integrant molecules "of several different substances, the molecules of one sub-"stance pass between those of the others, and each obey-"ing its own laws of attraction, crystallizes separately. "Besides the particular instances of this which occur in "our elaboratories, there is not a metalliferous vein which "does not furnish examples of it. Granite and a number "of the other aggregate rocks are striking examples of "this. Nature presents us with a very interesting fact in "substances confusedly crystallized, which contain dis-"seminated through them others in a state of regular and "perfect crystallization; such as compact granular lime-

"stone

"stone containing crystals of quartz, octahedral iron ore;
"and often other substances; such as masses of gyps which
"contain the same," (crystallized ?) "or crystals of bora-
"cite, of quartz, of carbonate of lime, or, lastly, of arrago-
"nite, as in Spain; such as masses of porphyry, &c. &c.
"In a solution which contains the integrant molecules of
"many different substances, but one of these in greater
"abundance than the rest, the cause which has prevented
"the crystallization of the last of these has had no effect
"on the others. Is it not natural to conclude, that, after
"the hasty and confused precipitation of the most abun-
"dant substance, the precipitated mass, being of a soft
"texture, has been penetrated with the solution loaded
"with the integrant molecules of the other substances
"which have afterwards united and been crystallized, as
"was the case with the alum and clay mentioned in § 64?
"This kind of union is known to mineralogists by the
"name of *elective attraction* or *affinity*. This explanation
"appears to me to be applicable to a great number of
"facts which we observe in nature; such, for example, as
"the formation of flint in nodules in chalk, or in larger
"masses in carbonate of lime." *Vide Traité de Mineralogie, par M. le Compte de Bournon*, 4to, 2me vol. p. 192. 193. § 70.

Note L. § 73. p. 129.

White Watson, F. L. S. in a small tract published in 1797, maintains that the mines in Derbyshire do traverse the toadstone. His account is as follows. "The rake "veins, containing ores in the incumbent limestone stra-"tum generally descend through the toadstone, but are "seldom found to contain ores when in the toadstone; ga-"lena (lead glance) is sometimes found with blende "and other minerals; but this circumstance rarely occurs: "the veins, when in the toadstone, are generally very "small, and composed chiefly of calcareous spar accompa-"nied with asphaltum; when they descend into the lime-"stone under the toadstone, they become thick again, and "frequently as rich in ores."

Note M. § 76. p. 141.

It would appear from later observations, that very considerable formations of lead glance occur in the coal formation in different parts of the county of Durham.

Note N. § 76. p. 142.

Werner remarks, "that in the substance of the primi-"tive rocks we never meet with the smallest trace of in-"flammable or coaly matter." Since the publication of this treatise, Werner himself and other mineralogists have discovered

discovered *slaty glance coal* in primitive mountains, in rocks of gneiss, mica slate, and clay slate.

NOTE O. § 125. p. 227.

Besides the formations of *lead glance* enumerated in the text, another may be added; it is that of Wanlockhead and Leadhills. At Wanlockhead the veinstones are *quartz*, *lamellar heavy spar*, *calc spar*, *brown spar*, and *mountain cork*. The ores are *lead glance*, *blende*, *manganese ochre*, *lead earth*, *sparry iron ore*, *calamine*, *brown iron ochre*, *iron pyrites*, *copper azure*, *green lead ore*, *white lead ore*, *lead vitriol* and *brown hematite*.

At Leadhills the veins are filled with the same materials as at Wanlockhead. These veins traverse *grey wacke* and *grey wacke slate*, and frequently contain fragments of both of these rocks.

At Strontian in Argyleshire, there is another formation of *lead glance*, which differs from that at Wanlockhead and Leadhills, and is probably of a different age from any of those enumerated in the text. It contains *lead glance*, *iron pyrites*, *calc spar*, *heavy spar*, *strontian*, *cross stone* and *foliated zeolite*.

Printed at the
Encyclopædia Britannica Press.

ERRATA.

page 6.	line 1.	*for* fossile	*read* fossil.
40.	— 14.	— beobachlungen	—— beobachtungen.
ibid.	— *ibid.*	— lagerstadle	—— lagerstätte.
page 45.	line 21	— storey	—— story.
63.	— 17.	*after* "lying"	*add* "side."
68.	— 8.	*for* fossile	*read* fossil.
75.	— 2.	*from bottom, for* composes	—— compose.

Made in United States
Troutdale, OR
07/21/2025

33080568R00169